Magnetic Nanoparticles for Medical Diagnostics

Series in Physics and Engineering in Medicine and Biology

About the Series

Series in Physics and Engineering in Medicine and Biology will allow IPEM to enhance its mission to 'advance physics and engineering applied to medicine and biology for the public good.'

Focusing on key areas including, but not limited to:

- clinical engineering
- diagnostic radiology
- informatics and computing
- magnetic resonance imaging
- nuclear medicine
- physiological measurement
- radiation protection
- radiotherapy
- rehabilitation engineering
- ultrasound and non-ionising radiation.

Magnetic Nanoparticles for Medical Diagnostics

Adarsh Sandhu
Graduate School of Information and Engineering, University of Electro-Communications, Tokyo, Japan

Hiroshi Handa
Department of Nanoparticle Translational Research, Tokyo Medical University, Japan

IOP Publishing, Bristol, UK

Multimedia content for this book is available from https://doi.org/10.1088/978-0-7503-1584-5.

ISBN 978-0-7503-1584-5 (ebook)
ISBN 978-0-7503-1582-1 (print)
ISBN 978-0-7503-1583-8 (mobi)

DOI 10.1088/978-0-7503-1584-5

Version: 20180801

IOP Expanding Physics
ISSN 2053-2563 (online)
ISSN 2054-7315 (print)

British Library Cataloguing-in-Publication Data: A catalogue record for this book is available from the British Library.

Published by IOP Publishing, wholly owned by The Institute of Physics, London

IOP Publishing, Temple Circus, Temple Way, Bristol, BS1 6HG, UK

US Office: IOP Publishing, Inc., 190 North Independence Mall West, Suite 601, Philadelphia, PA 19106, USA

Contents

Preface

This book was written to encourage members of the medical profession to join experts from other fields of research in exploiting the unique physical properties of magnetic nanoparticles for medical applications.

Reports on giant magnetoresistive (GMR) biosensors based on DNA hybridization and magnetic microbeads (originally intended for magnetic separation of biomolecules from aqueous solutions) to detect 'biological warfare agents' triggered global interdisciplinary research activities on the biomedical applications of magnetic nanoparticles [1]. Most of the initial research focused on the design and fabrication of GMR and Hall effect sensors to detect commercially available superparamagnetic microbeads functionalized with a wide range of complementary biomolecules such as avidin and biotin. This phase was led by physicists and electronics engineers with emphasis on the design of sensor arrays and for rapidly detecting many different targets simultaneously. Next, many more materials scientists, chemists and biologists joined this research and showed their prowess for synthesizing and functionalizing magnetic beads with diameters less than 500 nanometers. And importantly, in the last five years we have seen a dramatic increase in the participation of medical doctors and clinicians who are interested in applications including magnetic hyperthermia for the treatment of brain and prostate cancer, drug delivery, and 'magnetic staining' for rapid analysis of tissues during biopsies.

So this evolution from a small group of scientists fabricating magnetic sensors to truly multidisciplinary research on the wide ranging medical applications of magnetic nanoparticles illustrates the regenerative and dynamic nature of this area of research.

This book offers a cross-section of research being conducted by leading scientists based in Japan in 2017/2018. Some of the chapters include illustrative animation and videos to convey information.

The book covers the following topics:

- Magnetic probe and magnetic nanoparticles for sentinel lymph node biopsy;
- Magnetic separation of endosomes, exosomes, mitochondria, autophagosomes using magnetic beads;
- New affinity 200 nm diameter nanobead technology and target isolation of bioactive compounds;
- Fluorescent magnetic beads for medical diagnostics including the rapid detection of the prostate cancer biomarker PSA;
- Medical applications of functional viral capsid-coated artificial beads, including magnetic nanoparticles;
- Magnetic hyperthermia using implant type heating mediators;
- Synthesis, dispersion and application of stable magnetic colloids including MRI contrast agents and drug delivery;
- Optical and magnetic detection of fluorescent magnetic beads for medical diagnostics for point of care testing.

The contents will be updated with new chapters by colleagues who were not able to contribute at this time. The authors welcome enquiries and hope that the book will act as a catalyst to drive this area of research forward to solve real-life issues in medicine and medical sciences.

Adarsh Sandhu and Hiroshi Handa
Enquiries
Adarsh Sandhu UEC, Tokyo.
E-mail: sandhu@uec.ac.jp

Reference

[1] Baselt D R *et al* 1998 A biosensor based on magnetoresistance technology *Biosens. Bioelectron.* **13** 731–9

List of contributors

Associate Editor

Jaiyam Sharma
Graduate School of Information and Engineering,
University of Electro-Communications,
Tokyo,
Japan

Contributors

Masaki Sekino
Graduate School of Engineering,
The University of Tokyo,
Japan

Moriaki Kusakabe
Graduate School of Agricultural and Life Sciences,
The University of Tokyo,
Japan

Mari Takahashi
School of Materials Science,
Japan Advanced Institute of Science and Technology,
Japan

Shinya Maenosono
School of Materials Science,
Japan Advanced Institute of Science and Technology,
Japan

Takumi Ito
Department of Nanoparticle Translational Research,
Tokyo Medical University,
Japan
PRESTO, JST, Japan

Mamoru Hatakeyama
Tamagawa Seiki, Co, Ltd,
Japan

Satoshi Sakamoto
School of Life Science and Technology,
Tokyo Institute of Technology,
Japan

Maasaki Kawano
Department of Immunology,
School of Medicine,
Saitama Medical University,
Japan

Masanori Matsui
Department of Microbiology School of Medicine,
Saitama Medical University,
Japan

Takashi Nakagawa
Graduate School of Engineering,
Osaka University,
Japan

Balachandran Jeyadevan
Department of Material Science,
School of Engineering,
The University of Shiga Prefecture,
Japan

Kohki Noda
Wakayama University,
Japan

Ryoji Yukino
Graduate School of Information and Engineering,
University of Electro-Communications,
Tokyo,
Japan

Tsukasa Takamura
Toyohashi University of Technology,
Japan

Editor biographies

Adarsh Sandhu

Adarsh Sandhu is a Professor at the University of Electro-Communications, Tokyo (UEC, Tokyo), and Advisor to the President. He came to Japan in 1985 as a Monbusho Scholar to study at the Tokyo Institute of Technology and University of Tokyo. Then, in 1986, after completing his doctorate at the University of Manchester, UK, he joined Fujitsu Laboratories Ltd, Atsugi, focusing on gas source molecular beam epitaxy of AlGaAs/GaAs heterostructures. In 1992 he was a visiting scholar at the Cavendish Laboratory, and moved to Tokai University in April 1995, where he initiated research on magnetic imaging using scanning Hall probe microscopy. In August 2002 he accepted a tenured position at Tokyo Institute of Technology, and started work on biosensors based on magnetic labels for medical diagnosis. In 2010 he moved to Toyohashi University of Technology, as Vice Director of the research center 'EIIRIS', and in 2015 moved to his current position at UEC Tokyo in 2015. He is a visiting Professor at Tsinghua University (Beijing) and has collaborative projects with IT Delhi/ IISc, Bangalore/ UCL/ Pierre and Marie Curie University/ DGIST.

He is a member of Japanese academic societies, including the Japan Society of Applied Physics (JSAP), being an Executive Director from 2014 to 2016. He is also a member of the Foreign Correspondents Club of Japan (FCCJ).

Dr Sandhu is fluent in Japanese—spoken and written—giving lectures on solid-state physics, electromagnetism, x-ray diffraction and other courses, in Japanese. He enjoys most sports, astrophotography, and writing about scientific developments in the Asia Pacific region. His observations of the impact of science and technology on society in Asia form the backdrop to his first novel—a work in progress.

Hiroshi Handa

Hiroshi Handa graduated from Keio University, School of Medicine, in 1972 and received his MD and PhD in 1976 from the above. He was appointed as Assistant Professor at the University of Tokyo in 1976. He then worked as a postdoctoral fellow at Massachusetts Institute of Technology under the supervision of Professor Philip A Sharp from 1978 to 1980. He was appointed as Associate Professor in 1984 at the University of Tokyo, and as Professor in 1991 at Tokyo Institute of Technology. Since 2012, he has been a Professor at Tokyo Medical University and an Emeritus Professor at Tokyo Institute of Technology. He successfully completed several government-funded research projects as a project leader. His notable achievements are (1) discovery of the negative transcription elongation factors, DRB sensitivity-inducing factor

(DSIF) and negative elongation factor (NELF), which directly bind to RNA polymerase II (Pol II) and regulate the rate of Pol II-transcription elongation, including promoter proximal pausing, (2) the development of a one-step affinity purification technology using nano-sized latex and magnetic beads, which have been named Handa beads, (3) the discovery of a target of thalidomide teratogenicity, which led to an international research collaboration with Celgene in the US toward the development of novel drugs, (4) the development of highly functionalized nanobeads encapsulating both magnetite nanocrystals and fluorescent chemical compounds as high-speed and high-sensitive biosensor probes, and (5) the development of bio-nanocapsules formed by viral capsid proteins via self-assembled reconstruction for use as vaccine formulations.

Contributor biographies

Jaiyam Sharma

Jaiyam Sharma received a Bachelor of Technology in Engineering Physics from Indian Institute of Technology Delhi, India (IIT Delhi) in 2010. He received his Master of Engineering in Electrical and Electronic Engineering from Toyohashi University of Technology, Japan. His research interests are micro-electromechanical systems for medical diagnostics with a focus on the development of point of care diagnostics systems. He is currently a doctoral candidate at the Department of Engineering Science at The University of Electro-Communications, Tokyo. He is a nominated representative of the Japanese Society of Applied Physics (JSAP) and a member of the IEEE.

Masaki Sekino

Masaki Sekino received his PhD in engineering from the University of Tokyo in 2005, majoring in electronic engineering. His thesis was on novel techniques for mapping the electromagnetic properties of biological tissues using magnetic resonance imaging. He held a research associate position at the Graduate School of Medicine and at the Graduate School of Frontier Sciences. Since 2011, he has been an Assistant Professor of the Department of Electrical Engineering and Information Systems, Graduate School of Engineering, the University of Tokyo. His research interests are the biomedical applications of electromagnetic fields, such as magnetic resonance imaging, magnetic brain stimulation, and wearable electronic devices. He is now developing a handheld magnetic sensor for sentinel lymph node biopsy, and his device is used in multi-center clinical trials. He is an editorial board member of Scientific Reports.

Moriaki Kusakabe

Moriaki Kusakabe received his PhD from Nagoya University, Graduate School of Medicine, Doctor course, Department of Anatomy in 1984. His thesis was on 'Early development of mouse anterior pituitary: role of mesenchyme'. He held assistant and associate professor Nagoya University, School of Medicine, Department of Anatomy from 1982 to 1988. From 1989 to 2000, he was a research scientist in the laboratory of cell biology, Tsukuba Life Science Center, RIKEN. He became Director of Bio-resources Center from 1999. He founded Matrix Cell Research Institute Inc. in May 2005 and started the development of magnetic probe for sentinel lymph node biopsy. Since June 2007, he has been Project professor, Development of Advanced technology

Laboratory, Research Center for Food Safety, Graduate School of Agricultural and Life Sciences, The University of Tokyo. His research interests are cancer biology, the cancerous cells and surrounding mesenchymal cells interactions and bio-function of extracellular matrix protein Tenascin-C on carcinogenesis and cancer metastasis. He is a councilor of Japanese Society for Regenerative Medicine.

Mari Takahashi

Mari Takahashi was a PhD student at the Japan Advanced Institute of Science and Technology (JAIST). In April 2013, she started her master studies in JAIST under the supervision of Professor Shinya Maenosono working on the synthesis of magnetic-plasmonic hybrid nanoparticles for cellular organelles. In April 2015 she started her PhD studies in the same group to expand her master's work into the magnetic isolation of autophagosomes using magnetic-plasmonic hybrid nanoparticles. She currently works for Kao Corp.

Shinya Maenosono

Shinya Maenosono received his PhD in Chemical System Engineering from the University of Tokyo in 2002. He worked for Mitsubishi Chemical Corp. from 1993 to 1999 and for Fujifilm Co., Ltd from 1999 to 2000 as a researcher. In March 2000 he was appointed Assistant Professor of Chemical System Engineering at the University of Tokyo. Then, he moved to the Japan Advanced Institute of Science and Technology (JAIST) as an Associate Professor in 2006. At JAIST he leads his research group. His research in JAIST has focused on two main areas of interest in the field of materials chemistry and nanotechnology. The first area involves wet chemical synthesis of semiconductor nanoparticles with controlled size, shape and composition for energy conversion device applications. The second area focuses on the synthesis and bioapplication development of monometallic and alloyed multimetallic nanoparticles.

Takumi Ito

Takumi Ito was born in Hokkaido, Japan. He received a BSc in 2004, MEng in 2006, and PhD in Engineering in 2010 from the Graduate School of Bioscience and Biotechnology, Tokyo Institute of Technology. From 2006–2008 he was a JSPS research fellow. From 2008–2012 he was a researcher at the Integrated Research Institute of Tokyo Institute of Technology. From 2012–2013 he was an Assistant Professor of the Graduate School of Bioscience and Biotechnology at the Tokyo Institute of Technology. From 2013–2016 he was an Assistant Professor at the Department of Nanoparticle Translational Research of Tokyo Medical University. Currently, he is an Associate Professor at the same affiliation. He has also been a PRESTO researcher since 2015. His current research interests are the target isolation of chemical compounds and functional analysis of cereblon, a direct target of thalidomide. He has received the Tejima Memorial Research Award in 2011, Inoue prize for young scientists in 2012, and the Sassa Memorial Award in 2017.

Mamoru Hatakeyama

Mamoru Hatakeyama received his PhD in affinity purification by polymeric microspheres from the Department of Material Science, Faculty of Science and Technology, Graduated School of Keio University in 1998. He held a post-doc position for one year at the National Institute of Sericultural and Entomological Science consisting of researching entomological functions. Afterwards, he also held a post-doc position for four years at the Frontier Collaborative Research Center, Tokyo Institute of Technology, researching and developing polymer beads for the screening of target proteins. Thereafter he worked for Aphoenix Inc, and Sony Co. as a researcher. Since April 2006 he had been an Associate Professor of the medical and biotechnology project at Solutions Research Laboratory, Tokyo Institute of Technology. At the Tokyo Institute of Technology he had taken part in the applications of medicine and biotechnology by magnetic nanoparticles. From 2012 to 2015 he worked for Japan Stent Technology (JSTec) Co, Ltd as a General Manager in the R&D department. Since 2015 he has worked for Tamagawa Seiki Co. Ltd as an Assistant Manager at Biotronics Laboratory. He been researching and developing magnetic nanobeads. His research focuses on polymer chemistry and magnetic nanoparticles for applications of medicine and biotechnology.

Satoshi Sakamoto

Satoshi Sakamoto received a PhD in Chemistry from the Graduate School of Science, Tohoku University in 2003. He held a postdoctoral fellow position at the Department of Chemistry, University of Pennsylvania (2003–2005). Since August 2005 he has been an Assistant Professor at the Department of Biological Information, Graduate School of Bioscience and Biotechnology, Tokyo Institute of Technology (Tokyo Tech). In April 2016 he belonged to the School of Life Science and Technology, Tokyo Tech. His research interests are synthetic organic chemistry and chemical biology. His current research focuses on the construction of a novel sensing system using functionalized polymer-coated magnetic beads.

Maasaki Kawano

Masaaki Kawano received his PhD in technology from the Tokyo Institute of Technology Graduate School of Bioscience and Biotechnology in 2006. He held a post-doc position for three years at the University of California, Davis Department of Molecular and Cellular Biology. Then, in May 2010, he started an Assistant Professor position at Saitama Medical University. Since October 2016 he has been an Associate Professor of Saitama Medical University, Department of Allergy and Immunology Faculty of Medicine. He has been engaged in the construction of medical application technologies using the simian virus 40 virus-like particle composed of VP1 capsid for more than ten years from the beginning of his career. He is currently focused on the development of a novel adjuvant consisting of the simian virus 40 VP1 capsid for protecting virus infection and eliminating cancer from the body.

Masanori Matsui

Masanori Matsui received his PhD in Immunology from the Department of Allergology, Graduate School of Medicine, University of Tokyo, Japan in 1987. He held a post-doc position, followed by a research associate position at the Department of Microbiology and Immunology, The University of North Carolina at Chapel Hill, USA, studying MHC class I molecules and cytotoxic T lymphocytes (CTL). From April 2001 he has been an Associate Professor at the Department of Microbiology, Saitama Medical University, Japan. His research interests are the mechanism of MHC class I antigen processing and the development of a CTL-based vaccine against tumors and pathogenic viruses.

Takashi Nakagawa

Takashi Nakagawa worked at the Faculty of Engineering, Osaka University, as an Assistant Professor since June 1995 and received his PhD in engineering from Osaka University in 1999. From April 2004 he worked as an Associate Professor of the Department of Management of Industry and Technology, Graduate School of Engineering, Osaka University. He moved to the Graduate School of Science and Engineering, Tokyo Institute of Technology in 2006. He went back to Osaka University in 2009 and has been a Professor since 2018. His speciality is magnetic materials science and material analysis using radiation beams.

Balachandran Jeyadevan

Balachandran Jeyadevan is Professor at the Department of Material Science of The University of Shiga Prefecture, Japan. His expertise is in the synthesis of metallic and oxide nanostructures with different shapes and sizes such as nanoparticles, nanorods and nanowires using a polyol process for engineering application.

Kohki Noda

Kohki Noda graduated from the Electrical Engineering Department of Tokyo Metropolitan University in 1967. He received his PhD in Physical Electronics from the Tokyo Institute of Technology in 1998. He worked for IBM, both in Japan and overseas, until 2001, mainly in the area of hard disk drives. After that, he worked for The Institute of Physical and Chemical Research (RIKEN) in the area of protein-related analytical instruments. He is currently teaching a graduate class at Wakayama University. His research interests are sensing technology for allergen and antibody interactions, and its instrumentation. He is an IEEJ professional of IEEJ, a senior member of IEICE, a member of IEEE, MSJ, and JSAP.

Ryoji Yukino

Ryoji Yukino received a Bachelor and Masters degree in Electrical and Electronic Engineering from Toyohashi University of Technology. His research interests are the applications of guided mode resonance sensors for medical diagnostics with special emphasis on the development of point of care diagnostics systems. He is currently a doctoral candidate at the Department of Engineering Science, University of Electro-Communications, Tokyo. Ryoji has presented his research at several international conferences.

Tsukasa Takamura

Tsukasa Takamura received his PhD in nano-sized magnetic particles' detection from the Tokyo Institute of Technology in 2013. He held a tenure track Asistant Professor position for five years at Toyohashi University of Technology and collaborated with the Sandhu Laboratory for realizing 200 nm sized fluorescent magnetic particles' detection. His research interests are point of care testing, nano-sized magnetic particles' detection, magnetic sensors, and protocol for detecting bio-materials.

IOP Publishing

Magnetic Nanoparticles for Medical Diagnostics

Adarsh Sandhu and Hiroshi Handa

Chapter 1

Portable magnetic probe for detecting magnetic nanoparticles inside the body

Masaki Sekino and Moriaki Kusakabe

1.1 Sentinel lymph node biopsy

As breast cancer is the most common cancer among women in many countries, it is important to develop effective diagnostic and therapeutic methods. To select the appropriate therapeutic method for each breast cancer patient, the cancer stage must be correctly identified and the presence and extent of metastasis must be determined.

Metastasis refers to cancerous cells spreading to other organs through the lymphatic and vascular systems, where it subsequently develops. Breast cancer is known to metastasize through the lymphatic system. The lymphatic flow of the breast accumulates in the areola and lymph vessels extending from the areola to the axilla. Lymph nodes are present in the axilla. Thus, because of this structure, in many cases the cancer first metastasizes to the axillary lymph nodes, and subsequently to other organs.

Lymph nodes must be biopsied in order to determine the presence of metastasis. Hence, the lymph nodes are resected, and a prepared section of the biopsy sample is microscopically examined by a pathologist. Staining of the section with hematoxylin and eosin (HE) allows for the assessment of the cellular morphology, which in turn enables the pathologist to differentiate between healthy and cancerous cells. Alternatively, compared to HE staining, immunostaining of the section allows the cancer cells to be identified with higher precision. The properties and malignancy of cancer can also be identified.

The diagnosis of metastasis itself is an invasive process with side effects. If all the axillary lymph nodes are resected, lymphatic flow in the upper arm would be prevented, causing severe lymphedema, among other problems. Such side effects would significantly lower the patient's quality of life.

To minimize such side effects, the concept of sentinel lymph node biopsy (SLNB) is presently being introduced [1, 2]. Sentinel lymph node (SLN) refers to the lymph

doi:10.1088/978-0-7503-1584-5ch1

node that the cancer cells reach first when metastasizing via the lymphatic flow. Cancer first metastasizes to an SLN, then subsequently metastasizes to other lymph nodes and organs. If there is metastasis to an SLN, the possibility of subsequent metastasis to other organs must be assumed. However, if there is no metastasis to an SLN, one can assume that there is no metastasis to other lymph nodes or organs. In such a case, there is no need to resect the surrounding lymph nodes for examination, and these lymph nodes can be conserved. As a result of the advancement in breast cancer examination methods, breast cancer can presently be diagnosed at earlier stages. In developed countries, many cases are negative for SLN metastasis. Thus, SLNB enables the conservation of lymph nodes in many cases, and minimizes the side effects. Figure 1.1 shows the flow of SLNB.

The most important technique is to identify the SLN. To do so, a tracer is administered near a cancerous lesion. The tracer follows the lymphatic flow to reach a lymph node via the same pathway as the cancer cells. Therefore, if the lymph node that the tracer reaches first can be identified, this is designated the SLN. Regardless of the metastasis pattern, the tracer accumulates in the SLN. In surgical resection for breast cancer, the SLN is first identified with the patient under general anesthesia. The identified SLN is extracted, and cancer metastasis is promptly identified. To that end, a frozen section of lymph node is prepared and HE stained for pathological diagnosis. Using this method, metastasis can be identified within 30 min of lymph node removal. If the result is negative, only the primary lesion in the breast must be removed and lymph nodes apart from the SLN can be conserved. If the result is positive, the axillary lymph nodes are also resected in addition to removal of the primary lesion.

Presently, there are two standard tracers used to identify an SLN: radioisotope (RI) and dye [3–6]. RI is a radioactive drug that consists of tin colloid or technetium phytate. Radiation from a tracer is detected with a gamma ray detector. A tracer, subcutaneously injected into the areola, accumulates in the SLN over several hours or more. Therefore, a tracer is administered the day before surgery. An RI tracer can be detected from a several centimeter distance or more; thus, by passing a gamma ray detector over the skin, the SLN can be identified. This is an advantage of RI. By incising only the skin over the identified SLN, scarring on the skin can be minimized.

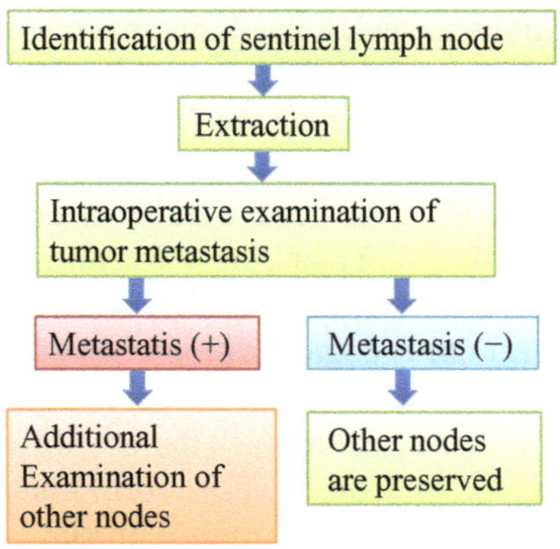

Figure 1.1. Flow of sentinel lymph node biopsy (SLNB).

In addition, because detection is made by a quantitative device output, judgment is objective. A gamma ray detector is used on the resected lymph node to record the count, which determines the amount of tracer that has accumulated in the lymph node.

There are two types of dye: blue dye, such as indigo carmine, and fluorescent indocyanine green (ICG). Accumulation of blue dye in a lymph node is identified with the naked eye, whereas the accumulation of ICG is identified using a dedicated infrared video camera. Because the molecular size of the dye is small, it can flow quickly through a lymph duct and accumulate in a lymph node soon after administration. Therefore, it is usually administered at the beginning of surgery after placing the patient under general anesthesia. Blue dye must be visually confirmed on the surface of the surgical field; however, with ICG, the video camera can visualize accumulation to a depth of several centimeters.

Depending on the individual SLN, both the RI and dye or either one alone will accumulate. By administering both RI and a dye, an SLN can be identified with greater certainty than by administering either one of the tracers. As using both tracers increases the rate of identification, guidelines in many countries, including Japan, recommend the use of both.

1.2 Use of magnetic nanoparticles for sentinel lymph node biopsy

As discussed above, the RI method uses a device for detection; thus, objective identification is possible. This method also has the advantage that an SLN can be detected even when lymph nodes are not exposed on the surface of the surgical field. However, a dedicated facility is necessary to use RI. Presently, many medical facilities do not have a dedicated RI facility; however, surgical treatment of breast cancer is provided in such facilities using only the dye method [7]. The identification rate decreases when only the dye method is used. If ICG is used as a dye, lymph nodes at depths of up to several centimeters can be identified. Because the dye tends to diffuse in the surgical field due to its high fluidity, identification can be difficult in some cases. In facilities where only the dye method is used, a change to use a detection device and a tracer with characteristics similar to those of RI is desirable.

As a technology that can meet such a need, magnetic nanoparticles (MNPs) have been used as a tracer in recent years [8–13]. As shown in figure 1.2, MNPs are injected near the lesion in the breast, and the MNPs that have accumulated in an axillary SLN are detected using a magnetic sensor. An example of MNPs, superparamagnetic iron oxide (SPIO) has been used as a contrast agent in magnetic resonance imaging (MRI) [14, 15]. Their safety for administration to humans has been established. Particles with a diameter of about 10 nm, in which a γ iron oxide core with diameter of several nm is coated with carboxydextran, form an aggregate with a mean particle diameter of 59 nm, which is called ferucarbotran. These particles are dispersed in aqueous solution. This is sold under the product name Resovist, containing 44.6 mg of iron in a 1.6 mL injectable solution. Another product, Sienna+, is a 2 mL injectable solution with similar particles in a different solvent. Sienna+ is manufactured for use in identifying SLNs [8].

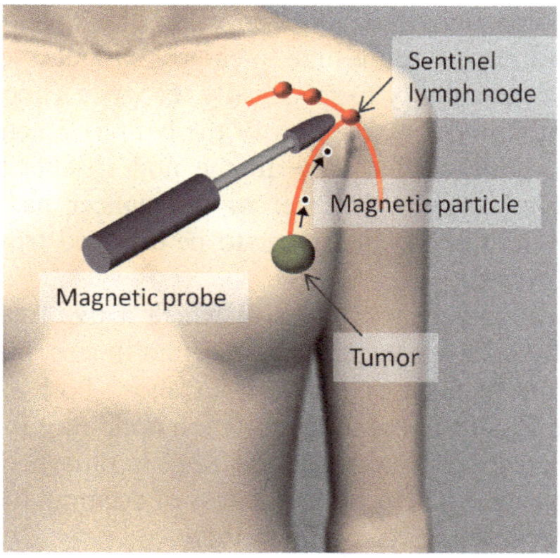

Figure 1.2. SLNB using magnetic nanoparticles (MNPs) [26].

Magnetic tracers can be used in medical facilities without the need for special RI facilities. Magnetic particles that are not exposed on the surgical field surface can be detected using a device, and the intensity of the signal can be quantified, thereby assuring objectivity. There is no problem with detection owing to rapid tracer diffusion. These MNPs are classified as a medical device in Europe, whereas they are classified as pharmaceuticals in Japan.

1.3 Pharmacokinetics of magnetic nanoparticles in the lymphatic system

With breast cancer, cancerous cells metastasize via the lymphatic system. When a tracer is injected into a cancerous tumor, it follows the same path as the cancerous cells and accumulates in the lymph nodes. However, when administering tracers to the tumor during surgery, it is necessary to accurately ascertain the location of the tumor and guide the tip of the tracer needle to it in three dimensions. In breast tissue, lymph vessels are concentrated around the areola and extend from there toward the axilla. As such, even if a tracer is administered subcutaneously to the areola, it travels to the axilla, making it possible to identify axillary SLNs. At present, breast cancer metastasis is usually diagnosed by administering a tracer to the areola, rather than the tumor.

The physician must determine the amount of time from administering the tracer to detecting its accumulation in the axillary lymph nodes when performing SLNBs in clinical practice. As such, the pharmacokinetics of the tracer need to be evaluated. The particles of dye, which are small and highly mobile, reach the lymph nodes within a few minutes of administration. However, RI tracers do not flow easily and take more than a few hours to reach the lymph nodes after administration. Accordingly, many medical institutions administer RI tracers the day before surgery. There are also individual differences in the dynamics of tracers, and accumulation tends to require more time in elderly patients.

Tracers do not have the ability to specifically recognize cancerous cells, and they are carried to the lymph nodes by the flow of the lymph. When cancer metastasizes to the lymph nodes, it may affect flow in the nodes. A tracer must be able to accumulate in the lymph nodes whether or not cancer has metastasized there. Furthermore, it is preferable for the tracer to be excreted from the body after an SLNB has been performed.

Figure 1.3 shows human lymph nodes in which a dye tracer and MNPs have accumulated. Lymph nodes are usually about 5 mm in size, but can grow to about 20 mm due to cancerous metastases. Figure 1.3(a) shows both a dye tracer (blue) and MNPs (dark brown). Figure 1.3(b) shows a lymph node slice for use in pathological diagnosis. Areas where MNPs accumulated appear blue because the tracer was stained using Prussian blue. There is a great deal of accumulation around the edges of the lymph node because the lymph flows there.

A vibrating sample magnetometer can be used to quantify the MNPs accumulated in the lymph node. With this device, samples, such as lymph nodes, are placed in a homogeneous magnetic field and magnetized. The magnitude of the magnetic moment produced by the magnetization is proportional to the amount of iron accumulated in the lymph nodes. The lymph node sample is made to vibrate at a set frequency. If a detection coil is placed near the sample, the sample's magnetic moment causes the magnetic flux to the coil to change over time, which generates AC voltage in the detection coil. The amount of accumulated iron can be calculated using the voltage. A gradiometer-type instrument is used in the detection coil, and a superconducting quantum interference device (SQUID) may also be used for higher sensitivity. Figure 1.4 shows the results of measurements taken for human lymph nodes using a vibrating sample magnetometer. The horizontal axis shows the external magnetic field applied to the sample, and the vertical axis shows the magnetic moment generated in the sample. Due to the properties of the MNPs, the magnetic moment increases linearly when the applied magnetic field is small, but when that field exceeds 50 mT, the increase in the magnetic moment

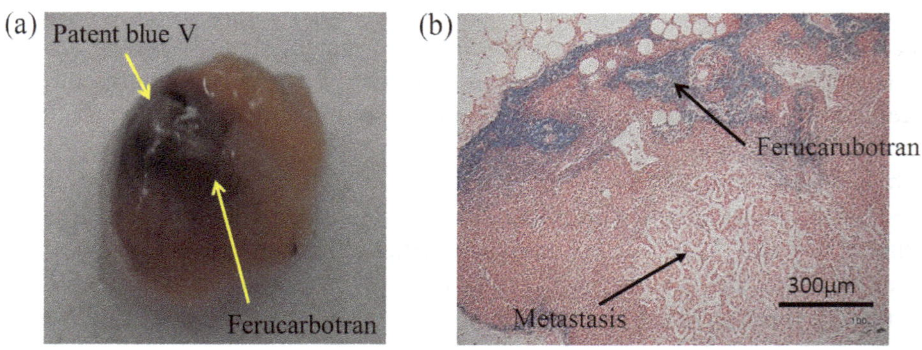

Figure 1.3. (a) Human sentinel lymph node (SLN) with accumulations of blue dye and MNPs (ferucarbotran). (b) Section of SLN with ferucarbotran stained with Prussian blue [16].

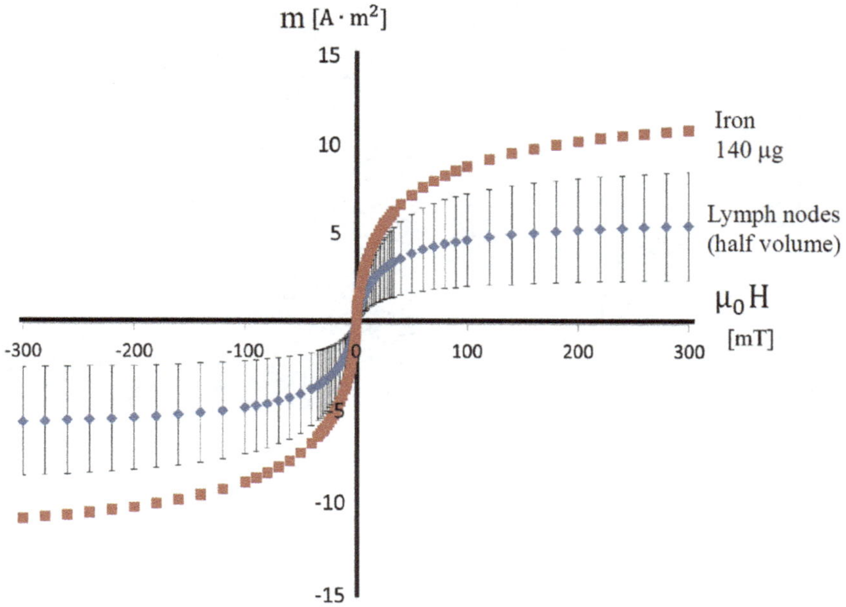

Figure 1.4. Quantification of iron accumulating in SLNs using a vibrating sample magnetometer [16].

reaches a limit at saturation. The figure shows the measurement results from a reference sample with 140 μg of iron and a lymph node cut in half. The error bars are the standard deviation of the measurements from six lymph nodes. These results suggest that approximately 140 μg of iron accumulates in an entire lymph node.

In recent years, scientists have been developing methods to apply an alternating magnetic field to samples instead of vibrating them; devices can be made smaller by eliminating moving parts [18]. As shown in figure 1.4, the subsequent changes in the magnetic moment deviate from the sine wave when an alternating magnetic field, which changes sinusoidally, is applied; this is because the magnetic moment of the MNPs has the property of saturation. It is possible to extract the signals originating from the MNPs if the harmonics are retrieved through signal processing. Biological tissue is diamagnetic, and it does not produce harmonics because its magnetic moment is proportional to the strength of the external magnetic field. As such, it is advantageous to use alternating magnetic field methods because they are not affected by the diamagnetism of biological tissue.

Animal study is useful for systematically clarifying the pharmacokinetics of MNPs. This is especially true because it allows researchers to optimize the time from administration to detection, and study whether the most appropriate administration site is within the tumor or subcutaneously in the surrounding area. For instance, researchers have published papers in which MNPs were injected into the legs of mice, and the accumulation in popliteal lymph nodes and inguinal lymph nodes was evaluated [19]. As shown in figure 1.5(a), cancerous cells (B16–BL6) were injected into the right leg, and skin cancer developed over a one-week period. A 5 μL solution

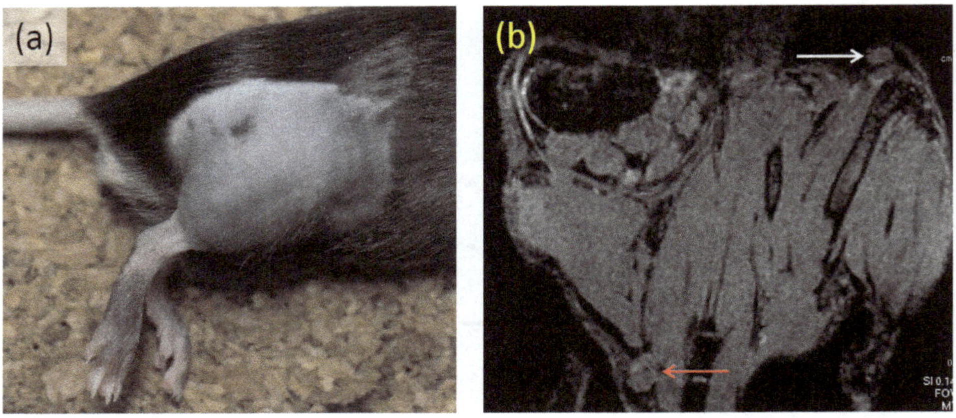

Figure 1.5. (a) Rat model at seven days after introducing cancer cells to right lower limb. (b) Pre-injection T1-weighted magnetic resonance imaging (MRI) with white and red arrows indicating a left inguinal and right popliteal node respectively [19].

of MNPs (Sienna+) and a dye (Patent Blue) were injected intratumorally into the right leg and subcutaneously into the left leg. Following this, an MRI and lymph node excision were performed immediately after administration, and 1, 4, and 24 h later. In MRIs, T1- and T2-weighted images were taken using a device with a 7 T main field. As shown in figure 1.5(b), artifacts caused by local distortions of the magnetic field appeared in lymph nodes where Sienna+ accumulated; the size of these artifacts was evaluated. Additionally, the amount of accumulated iron in the excised lymph nodes was measured with a vibrating sample magnetometer, and slices were created for iron staining. Artifacts in the lymph nodes increased more quickly with subcutaneous administration than with intratumoral administration, according to the results gleaned from MRI observation over time. Additionally, as shown in figure 1.6, the results from measuring the lymph nodes demonstrated that the amount of accumulated iron increased as time passed following administration. These results reveal that, although accumulation to the lymph nodes occurs for either administration site, intratumoral administration requires physicians to wait for a longer time before conducting detection.

The pharmacokinetics of MNPs are also affected by particle size. A study has been published comparing the pharmacokinetics of particles of different sizes (diameters of 32 nm, 59 nm, and 111 nm) [17]. MNPs were injected into the areolas of pigs, and accumulation in the inguinal lymph nodes was evaluated. The accumulation was measured using two devices: a handheld magnetometer and a vibrating sample magnetometer. The results showed that the 59 nm MNPs accumulated in the lymph nodes were the quickest and in the largest amounts. This means, of the three particle sizes studied, 59 nm particles are the most appropriate for SLNBs. The previously mentioned Resovist and Sienna+ particles both have a diameter of 59 nm.

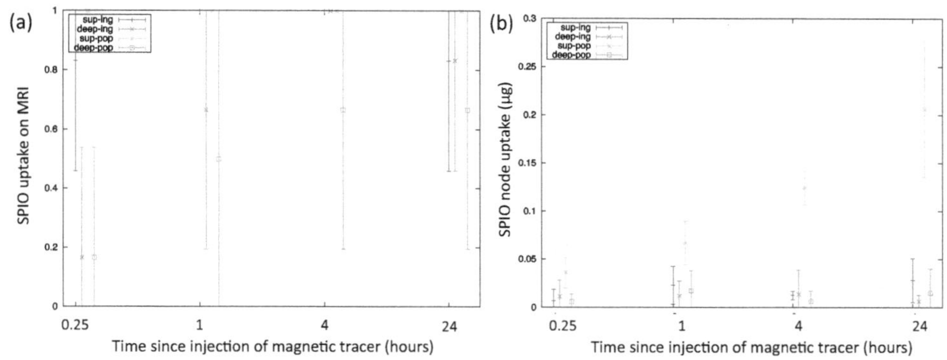

Figure 1.6. (a) Nodal uptake of MNPs recorded on MRI. (b) Nodal uptake recorded using a vibrating sample magnetometer [19].

1.4 Techniques for detecting magnetic nanoparticles

Resovist, a commercialized MNP, was developed as an MRI contrast agent. Resovist produces a negative contrast and creates signal reduction when it accumulates. MNPs concentrated in lymph nodes also exhibit a contrast effect and, as shown in figure 1.5(b), generate artifacts that make lymph nodes appear as dark voids on images. By comparing lymph node images before and after contrast, one can identify the lymph nodes where MNPs have concentrated [20]. As described, the technique of using MNPs and MRI to make lymph ducts and nodes visible is called magnetic resonance (MR) lymphography. As discussed previously, when approximately 1 mL of a MNP solution is administered, the MNPs accumulate in the SLNs whether or not cancer has metastasized to the lymph nodes. However, it has been reported that when a small amount of MNPs is diluted and administered, they only accumulate if there is no metastasis to the lymph nodes and do not accumulate if there is metastasis [21]. This is because when metastatic cancer develops inside a lymph node, the flow of lymph is hindered. Using this technique makes it possible to diagnose metastasis.

An MRI obtains signals from the hydrogen nuclei contained in biological tissue; the MNPs themselves do not create MRI signals. Meanwhile, a technology currently being developed, magnetic particle imaging (MPI), obtains signals from the MNPs themselves and makes their distribution visible [22, 23]. The magnetic properties of the MNPs are used to take measurements in MPI. The magnetic susceptibility of a tissue containing MNPs is defined as the rate by which magnetization rises when the external magnetic field is increased. As demonstrated in figure 1.4, because there is nonlinearity in the magnetization of MNPs, the magnetic susceptibility depends on the strength of the external magnetic field. The weaker the external magnetic field, the larger the magnetic susceptibility. Coils are positioned around the body, and a distribution of a static magnetic field is generated inside the body. At this time, the electric current to the coils is adjusted so that a point is created within the body where the magnetic flux density is zero. When MNPs are distributed within the body, the magnetic susceptibility of the tissue rises locally at the point at which the magnetic flux density of the static magnetic field is zero. If an alternating magnetic field is applied externally and the alternating magnetic field generated by the MNPs is detected, the magnitude of magnetization can be measured. This magnetization is the local value at the point at which the magnetic flux density of the static magnetic field is zero. If the system changes the electrical current flowing to the coils and moves the point at which the magnetic flux density is zero, it is possible to obtain the magnitude of magnetization at each point. Using such principles, the distribution of MNPs within the body can be visualized. The potential applications of MPI for identifying SLNs are being investigated by some research groups.

MRI and MPI require large equipment, making them difficult to use in operating rooms during surgery. Because the patient's posture during surgery is different from that during MRI and MPI imaging, the position of the lymph nodes also changes. Therefore, MRI or MPI alone are not enough; a small, easy-to-use device is needed to identify SLNs in the operating room. Therefore, researchers have been studying techniques for detecting MNPs with small magnetic sensors. Unlike equipment that uses radiation, a handheld magnetic probe can be introduced in regular operating rooms. Because the number of MNPs that accumulate in the lymph nodes is small, the magnetic sensor needs to be highly sensitive.

A SQUID is a magnetic sensor that offers extremely high sensitivity. Research on identifying SLNs with SQUIDs has been conducted at various research institutions [24]. MNPs are magnetized by geomagnetism, and the magnetic field generated by that magnetization can be detected using SQUIDs. Although SQUIDs have the major advantage of being highly sensitive, they require a coolant such as liquid helium or liquid nitrogen to achieve superconductivity. Additionally, they are strongly influenced by surrounding equipment due to their hypersensitivity, and it is difficult to distinguish between the magnetic field generated by the MNPs and the magnetic field generated by the surrounding equipment. Moreover, because the sensor must be placed inside an insulated container that stores the coolant, there are limits to miniaturization. For these reasons, SQUIDs have not become widespread in clinical practice.

In recent years, there has been rapid development of equipment using magnetic sensors that can be operated at room temperature. Magnetic sensors that operate at room temperature do not require coolant, making it possible for the sensor unit to be miniaturized to a handheld size. Because magnetic sensors that operate at room temperature do not have the high sensitivity of a SQUID, it is necessary to magnetize the MNPs and generate a strong, detectable magnetic field. In many cases, handheld sensor units are equipped with a magnetic field source inside. There are two methods used as magnetic field sources: using coils to generate an alternating magnetic field and using a permanent magnet to generate a static magnetic field. The advantage of coils is that they extract only the signal of the operating frequency, preventing them from being affected by static magnetic fields, such as geomagnetism. However, they also present some issues, such as the time required to stabilize after startup because of heat generation and the larger drive circuits needed due to energy consumption. One product that uses an alternating magnetic field in a handheld sensor is Sentimag, which has been approved in Europe and the United States.

1.5 Portable magnetic probe using a permanent magnet

To identify SLNs using a handheld magnetic probe, the physician searches for lymph nodes while scanning the surgical field with the magnetic probe. Making magnetic probes smaller considerably improves their operability. Additionally, using a permanent magnet as the source of the magnetic field makes it possible to efficiently produce a strong magnetic field without any accompanying electric power consumption or heat generation. Accordingly, magnetic probes using permanent magnets are a good fit for miniaturization, making them useful within the surgical field for searching lymph nodes. Because permanent magnets do not produce heat, they are able to reduce the effects of temperature changes on the sensor. Additionally, because no electric power is needed to generate a magnetic field, drive circuits are smaller and portability is better.

When a magnetic probe is made of a handheld sensor and a desktop drive circuit, the magnetic probe is used within a sterilized environment and the drive circuit is placed outside the environment. This makes magnetic probes difficult to operate. When electric power consumption is kept low by using a permanent magnet, it becomes possible to enclose the entire battery and drive circuit within the handgrip. If such a design is adopted, the magnetic probe can be operated entirely within the sterilized area during surgery, vastly improving its operability.

By making full use of these characteristics, we fabricated a prototype magnetic probe that uses a permanent magnet and small magnetic sensor that can be operated at room temperature. This probe was then used in clinical research [16, 25–28].

Figure 1.7(a) shows the structure of the head of the magnetic probe. The small magnetic sensor, which identifies magnetic fields generated by MNPs, is placed on the tip of the shaft. The magnetic field generated by the permanent magnet installed

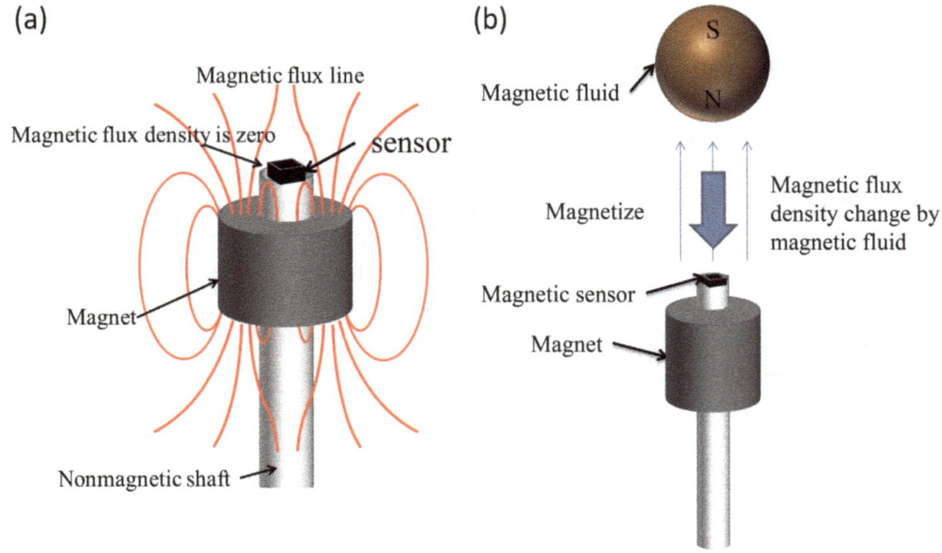

Figure 1.7. (a) Structure of a magnetic probe head equipped with a permanent magnet and Hall effect sensor. (b) Principle of a magnetic probe.

under the magnetic sensor magnetizes MNPs. Using nonmagnetic brass for the shaft prevents any influence on the measurements. The sensitive direction of the magnetic sensors and the magnetization direction of the permanent magnet are the same as the axial direction of the nonmagnetic shaft. Figure 1.7(b) shows the principles governing the way a magnetic probe detects MNPs. When an SLN where MNPs have accumulated comes close to the head, the MNPs are magnetized by the magnetic field generated by the permanent magnet. The magnetization of the MNPs generates a magnetic field in the surrounding area. As a result, the magnetic flux density applied to the magnetic sensor changes. Detecting this change makes it possible to identify the accumulation of MNPs. As shown in figure 1.7(a), there is a point on the central axis of the permanent magnet where the magnetic flux density becomes zero. The magnetic sensor is placed at this point. When there are no MNPs, the output of the magnetic sensor is zero; in other words, sensor offset can be eliminated. These strategies allow the sensor output signals to be amplified at a high gain, and enable minute changes in magnetic flux density to be detected. Hall effect sensors are suitable as small magnetic sensors. In the area around the point where the magnetic flux density becomes zero, the gradient of the magnetic field is extremely strong. Aside from Hall effect sensors, several types of high-sensitivity magnetic sensors using ferromagnetic materials have been put into practical use; however, these are difficult to operate under a strong magnetic gradient.

To identify lymph nodes from above the skin, it is preferable to design a large permanent magnet and generate a magnetic field to an increased distance from the magnet. However, the larger the magnet becomes, the larger the tip of the probe must be, leading to poorer operability. This trade-off between magnet size and operability is a challenge, and there is a demand for a magnet designed in a way that optimizes this.

Numerical analysis of the electromagnetic field using the finite element method was performed to evaluate the magnetic flux density applied to the sensor by the MNPs and the effect of magnet size on sensitivity. The point at which the magnetic flux density generated by the permanent magnet becomes zero was used as the reference point. A numerical model of lymph nodes that include MNPs was placed in the area around the sensor, and the value of magnetic flux density generated at the reference point was evaluated as sensitivity ΔB. The larger the value of ΔB, the higher the sensor output that can be obtained. When ΔB is larger than the variation in sensor output caused by noise or other interference, detection of MNPs is possible.

The coercive force of the permanent magnet was set to 976 kA m^{-1}, the inner radius to 2.5 mm, the outer radius to 6.25 mm, and the length to 12 mm. Figures 1.8(a) and (b) show the B_z component of the magnetic flux density generated by the permanent magnet when MNPs are not present. The range where the magnetic flux density exceeds 50 mT was up to 10.4 mm away from the reference point on the central axis. The gradient of the magnetic flux density in close proximity to the reference point was approximately 150 mT mm^{-1}.

Additionally, to investigate the distribution of the magnetic field arising from the MNPs, when magnetization was saturated, the permanent magnet was removed; a

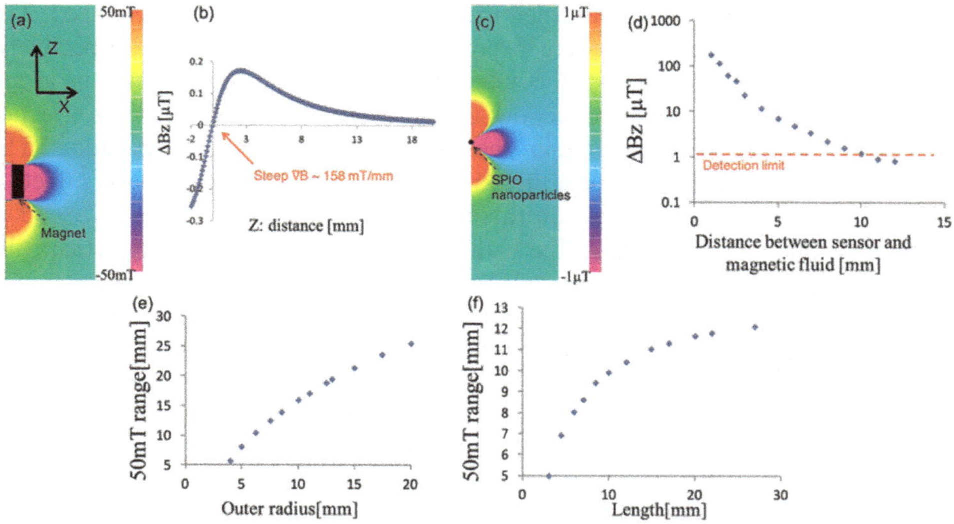

Figure 1.8. (a) Distribution of the longitudinal magnetic field B_z of the permanent magnet. (b) B_z of the magnet on the z-axis. (c) Distribution of B_z of 140 μg of fully magnetized MNPs (5 μL of Resovist) located at $z = 10.4$ mm. (d) Magnetic field strength ΔB at the sensor position as a function of the distance to the MNPs. Relationships between (e) the outer radius and (f) the length of the magnet, and the distance of the point where the B_z of the magnet is 50 mT [16].

numerical analysis of the magnetic field distribution was then conducted only for the magnetized MNPs. The MNPs were placed at a point 5 mm away from the reference point. Figure 1.8(c) shows the B_z composition of the magnetic field generated by the MNPs when a 2×10^3 A m^{-1} magnetization (equivalent to saturation) was applied. A magnetic flux density on the order of 1 μT reaches the area of the reference point.

These data were used to derive ΔB depending on the position of 5 μL of the MNPs. In figure 1.8(d) the horizontal axis shows the distance from the reference point to the MNPs, while the vertical axis shows ΔB at the reference point. The shorter the distance to the MNP, the larger the change in magnetic flux density. When the practical sensitivity limit of the Hall sensor is 1 μT, 5 μL of MNPs can be detected up to approximately 10 mm.

The effect of magnet size on sensitivity was evaluated thereafter. Given a distance between the MNPs and the sensor, the magnetic field, received by the sensor, was at its maximum when the MNPs were saturated. For this reason, we focused on the magnetic flux density at which the MNPs are saturated. Because MNPs are saturated at approximately 50 mT, according to the magnetic characteristics of Resovist shown in figure 1.4, we obtained the distance from the reference point to the 50 mT point. First, the inner radius and the length of the magnet were fixed at 2.5 mm and 12 mm, respectively, and only the outer radius was changed. Figure 1.8(e) shows the location on the central axis where magnetic flux density reaches 50 mT. The horizontal axis shows the outer radius of the permanent magnet. The larger the outer radius, the farther the magnetic field reaches, and the larger the range of saturation of the magnetization of MNPs becomes.

Figure 1.8(f) shows the results when the length of the magnet was changed. The inner radius and outer radius were fixed at 2.5 mm and 6.25 mm, respectively. Similar to the results obtained for the outer radius, the longer the magnet, the larger the range within which magnetic flux density exceeded 50 mT became. The effect of changing the length was smaller than the effect of changing the outer radius. The results revealed that when the length exceeds approximately 12 mm, the effect of lengthening the magnet becomes saturated.

The experimental sensitivity of the Hall effect sensor used in the magnetic probe was approximately 1 μT. A distance of 11 mm was found to be the distance at which 5 μL of MNPs with saturated magnetization generated 1 μT of magnetic flux density. Accordingly, it is preferable to design the probe such that approximately 11 mm is the range at which a magnetic flux density of 50 mT or more is generated. Based on figures 1.8(e) and (f), a magnet with an inner radius of 2.5 mm, outer radius of 6.25 mm, and a length of 12 mm fulfills nearly all these requirements. Accordingly, we used a permanent magnet of this size in our test-manufactured magnetic probe.

Furthermore, we created a head for the magnetic probe and conducted an experiment to evaluate its sensitivity. A ring-shaped neodymium magnet and a Hall effect sensor were installed at the end of a brass shaft. We changed the relative positions of the magnet and the Hall effect sensor in order to find the point at which the magnetic flux density was zero, measuring the output of the sensor at those positions. Magnetic flux density changes rapidly just around the point where the magnetic flux density produced by a permanent magnet is zero; therefore, it is necessary to determine a precise location for both the Hall effect sensor and the permanent magnet. Therefore, we developed a specialized positioning device that is adjustable on the micrometer scale, and used it to determine the location of the permanent magnet. The Hall effect sensor was implemented on a shaft, and this was attached to the positioning device. The location of the permanent magnet can be adjusted forward or backward on the order of micrometers by turning a handle. A laser displacement gauge can be used to measure the difference in position between the tip of the Hall effect sensor and the tip of the permanent magnet.

The permanent magnet was fixed to the position where the output (the magnetic flux density) of the Hall effect sensor would become zero. Once those positions were fixed, the distance from the Hall effect sensor to the MNPs was varied. At this point, the output of the Hall effect sensor was measured and compared to the results of the numerical analysis. The samples were prepared by putting 2 μL, 5 μL, 10 μL, and 20 μL of Resovist, corresponding to 56 μg, 140 μg, 280 μg, and 560 μg of irons, into resin vessels.

Additionally, we evaluated the optimal relative distance between the Hall effect sensor and the permanent magnet in order to detect the MNPs. We measured the changes in the output voltage with and without a sample containing 5 μL of Resovist. These measurements were repeated while changing the position of the permanent magnet in a stepwise manner. First, the position of the permanent magnet was shifted, and the output of the Hall effect sensor was then recorded. Thereafter, the Hall effect sensor was brought into contact with the sample, and the output was recorded. The differences in these output readings were investigated for different positions of the permanent magnet.

We changed the relative position of the sensor and the permanent magnet, and measured the output of the sensor when the MNPs were brought into contact with the tip of the sensors at various distances. Figure 1.9(a) shows the changes in the output of the sensor. The horizontal axis shows the distance between the permanent magnet and the sensor. The vertical axis shows the differences in the output of the sensor during contact with the MNPs. It can be seen that the change in sensor output was highest at the 0.3 mm point. Positioning the magnet near the point where the magnetic flux density was closest to zero resulted in the greatest changes in the magnetic flux density detected by the sensor. The range for obtaining maximum sensitivity was in the range of approximately ±0.1 mm from the point where the magnetic flux density was zero.

Figure 1.9(b) shows the changes in magnetic flux density when the sensor is brought close to the MNPs with the Hall effect sensor fixed in place at the point where the magnetic flux is zero. The horizontal axis shows the distance along the central axis from the tip of the sensor to the MNP sample. The vertical axis shows the magnetic flux density detected by the sensor. It can be seen that the magnetic flux density increased as the volume of Resovist increased. These results were similar to the results obtained through numerical analysis, considering the thickness of the sensor. Assuming that the sensor has a sensitivity within 1 μT, it can be used to detect 56 μg of iron at a distance of 7 mm, 140 μg at 9 mm, 280 μg at 11 mm, and 560 μg at 15 mm.

We fabricated a prototype magnetic probe head for use in SLNB in clinical study. The fixed position of the magnet is the point at which the sensor's magnetic flux density output is zero. The magnet, shaft, and sensor were fixed to each other and then detached from the positioning device. Figure 1.10(a) shows the structure of the

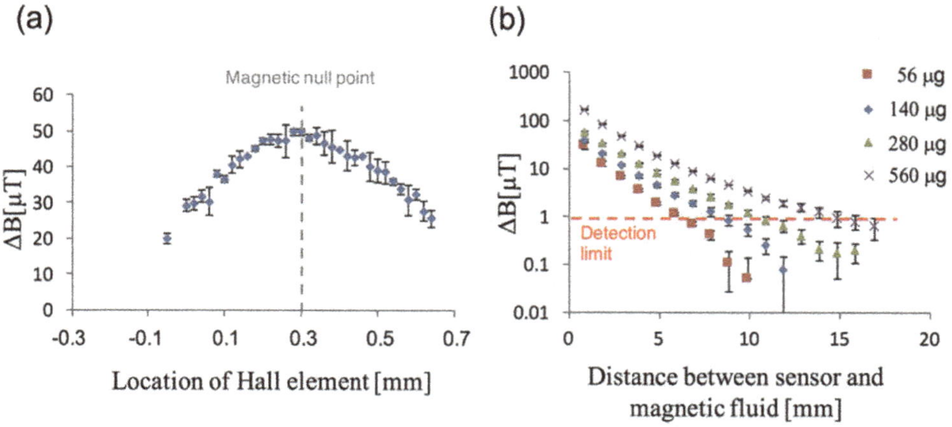

Figure 1.9. (a) Signals detected with 140 μg of iron (5 μL of Resovist), as functions of the Hall sensor position: the gray dotted line shows that the magnetic null point is at approximately 0.3 mm. (b) Measured signals for four different volumes, 56, 140, 280, and 560 μg of iron. The red dotted line shows the detection limit of the sensor [16].

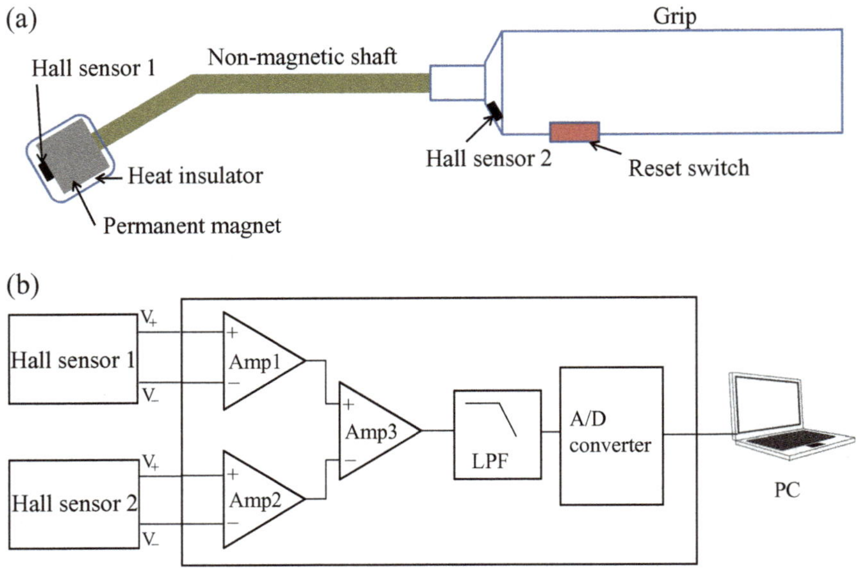

Figure 1.10. (a) Structure of the magnetic probe. (b) Diagram of the driving circuit.

magnetic probe. To improve the ease of use, the shaft was given a 30° curvature. Another Hall effect sensor was placed in the grip, and the geomagnetic field was measured. By taking the difference between this sensor and the head sensor, we canceled the effect of geomagnetism. These sensors were placed such that they were parallel to each other. A heat shield made with insulation paper was applied to the area around the MNP sensor on the head in order to reduce temperature changes when the probe comes into contact with the body.

As shown in figure 1.10(b), in the drive circuit, outputs from the two sensors are amplified and converted to a digital signal with an AD converter. Because the detected magnetic field is weak, the signal is easily affected by drift. A button to reset the drift is provided, and drift can be eliminated during operation by using the reset button as needed.

Figure 1.11 shows an updated prototype magnetic probe. The drive circuit and battery are housed inside the grip, and the device is designed to fit entirely in the hand. Magnetic flux density is shown on a small display in units of microtesla. A speaker is built in, and changes in the sound notify the user when MNPs are detected. Equipping the device with a speaker enables physicians to know when MNPs are detected, without having to check a screen during surgery. The sound output is a continuous pulse that changes in frequency, pulse length, and pulse interval according to the output of the magnetic sensor.

1.6 Clinical applications

Identification of SLNs in breast cancer patients was performed using a magnetic probe. There were six patients with an average age of 69 ± 7. After administering a general anesthesia, 1.6 mL of a MNP solution (Resovist) and a blue dye (1% patent blue, 3 mL) were injected subcutaneously to the areola. Figure 1.12(a) shows how these injections were administered. To ensure sufficient flow of MNPs and dye into the lymph ducts, the breasts were massaged for several minutes. After making an incision at the axilla, the magnetic probe was used to search for SLNs. Then, as

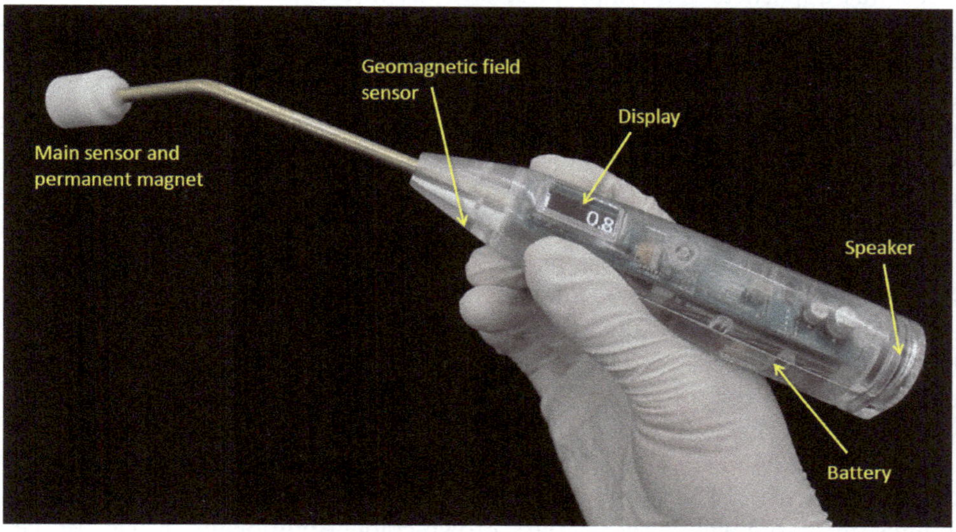

Figure 1.11. Prototype magnetic probe.

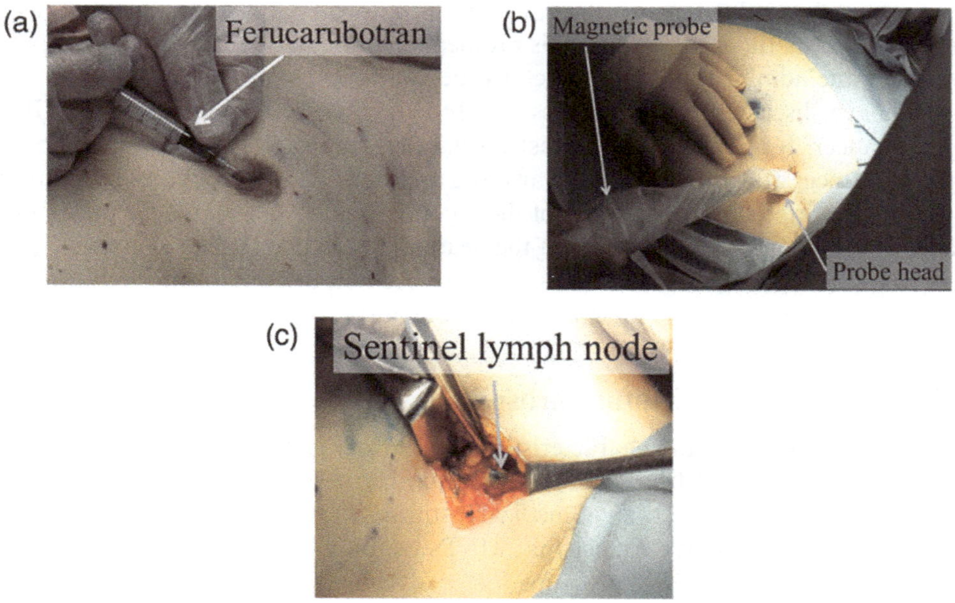

Figure 1.12. Clinical test using the handheld magnetic probe; (a) injection of magnetic tracers around the subareolar region, (b) transcutaneous detection of the SLNs before incision, and (c) SLNs detection and excision in the axilla after incision. In the injection phase (a), the blue dye tracer is injected at the same location after the injection of the magnetic tracer [16].

shown in figure 1.12(b), the tip of the magnetic probe was inserted into the surgical field. SLNs were identified by combining visual observation of dye and detection of MNPs, as shown in figure 1.12(c); the identified lymph nodes were excised. To confirm whether MNPs had accumulated in an excised lymph node, the magnetic probe was applied to the lymph node and the measured value of magnetic flux density was recorded. Because there are multiple SLNs in many cases, we explored the axial area using the magnetic probe to confirm that there was no remaining magnetic field response. To determine whether cancer had metastasized to the SLNs, rapid diagnoses were made during surgery using HE staining of frozen sections. Additionally, to quantify the number of MNPs accumulated in the excised lymph nodes, a vibrating sample magnetometer was used to measure the magnetic moment.

In four out of the six patients, accumulation of the MNPs in the SLNs was confirmed and detection with the magnetic probe was successful. Metastasis in the SLNs was observed in one patient. The SLNs where metastasis was observed were among those nodes successfully detected by the magnetic probe. The magnetic flux density value observed using the magnetic probe in the MNPs was 11 ± 5 μT. A number of MNPs proportional to 5 ± 3 μL of solution was contained in the excised lymph nodes.

Some of the MNPs administered subcutaneously to the areola remained in the area. Because the MNPs are a dark brown color, it is also possible that the administration site will develop a dark brown color. Additionally, it has also been reported that residual MNPs can create MRI artifacts. The presence of residual MNPs in patients has been found after periods of more than a year [29]. MRI artifacts such as these may affect the monitoring of patient progress following surgery. One method of avoiding this problem is to use MNPs only in patients who will have the administration site, such as the areola, excised.

We have discussed the usefulness of MNPs primarily in relation to SLNB of breast cancer. It is known that metastasis via the lymphatic system occurs in other types of cancer affecting various organs; the usefulness of SLNB in these conditions has been reported. Some examples of these conditions include stomach cancer, skin cancer, uterine cancer, and cancer of the head and neck. Regarding skin cancer, the usefulness of SLNB using MNPs has already been reported [30]. In the future, we expect further applications in other forms of cancer.

1.7 Conclusions and future prospects

This chapter describes the characteristics and current state of techniques for using MNPs as a tracer in SLNB. Because the number of people who develop breast cancer is high, it is treated even at medical institutions that do not have facilities that administer RI tracers. We expect that MNPs could replace RI as a tracer in such medical institutions. MNPs are superior to dye in that they can be objectively detected using a magnetic probe and are detectable even at a depth of approximately 1 cm.

Although a variety of devices are reported as being capable of detecting MNPs, a handheld magnetic probe can be used in the operating room. Devices using alternating magnetic fields have already been applied in practical settings. Because little energy is consumed when using a static magnetic field generated by a permanent magnet, entire electronics, including the drive circuit and battery, can be housed within the grip.

One major challenge moving forward is to make the sensor more sensitive so as to increase the detection distance of the magnetic probe. In recent years, the sensitivity of magnetic sensors that operate at room temperature has increased rapidly. For example, there have been reports on the development of magnetic probes using magnetic tunnel junction devices [31]. Additionally, fluxgate sensors, optical pumping magnetometers, and diamond nitrogen-vacancy (NV) magnetic sensors operate without coolant and are highly sensitive, which means that they may be applied to the detection of MNPs.

References

[1] Veronesi U *et al* 1997 *Lancet* **394** 1864–7

[2] Noguchi M, Miwa K, Michigishi T, Yokoyama K, Nishijima H, Takanaka T, Kawashima H, Nakamura S, Kanno H and Nonomura A 1997 *Breast Cancer* **4** 143

[3] Kim T, Giuliano A E and Lyman G H 2006 *Cancer* **106** 4–16

[4] Wetzig N, Gill P G, Espinoza D, Mister R, Stockler M R, Gebski V J, Ung O A, Campbell I and Simes J 2017 *Ann. Surg. Oncol.* **24** 1064–70

[5] Radovanovic Z, Golubovic A, Plzak A, Stojiljkovic B and Radovanovic D 2004 *Eur. J. Surg. Oncol.* **30** 913–7

[6] Hung W K, Chan C M, Ying M, Chong S F, Mak K L and Yip A W 2005 *Br. J. Surg.* **92** 1494–7

[7] Meyer-Rochow G Y, Martin R C and Harman C R 2003 *ANZ J. Surg.* **73** 815–8

[8] Douek M 2014 *Ann. Surg. Oncol.* **21** 1237–45

[9] Rubio I T, Diaz-Botero S, Esgueva A, Rodriguez R, Cortadellas T, Cordoba O and Espinosa-Bravo M 2015 *Eur. J. Surg. Oncol.* **41** 46–51

[10] Thill M, Kurylcio A, Welter R, van Haasteren V, Grosse B, Berclaz G, Polkowski W and Hauser N 2014 *The Breast* **23** 175–9

[11] Ahmed M, Purushotham A D and Douek M 2014 *Lancet Oncol.* **15** e351–362

[12] Ahmed M and Douek M 2013 *Biomed. Res. Int.* **2013** 281230

[13] Zada A *et al* 2017 *Br. J. Surg.* **103** 1409–19

[14] Fujita M *et al* 1996 *J. Magn. Reson. Imaging* **6** 472–7

[15] Kato N, Takahashi M, Tsuji T, Ihara S, Brautigam M and Miyazawa T 1999 *Invest. Radiol.* **34** 551–7

[16] Sekino M *et al* 2018 *Sci. Rep.* **8** 1195

[17] Pouw J J, Ahmed M, Anninga B, Schuurman K, Pinder S E, Hemelrijck M V, Pankhurst Q A, Douek M and ten Haken B 2015 *Int. J. Nanomedicine* **10** 1235–43

[18] Kuwahata A, Kaneko M, Chikaki S, Kusakabe M and Sekino M 2018 *AIP Adv.* **8** 056713

[19] Ahmed M *et al* 2016 *Nanomed. Nanotechnol. Biol. Med.* **12** 1045–52

[20] Shiozawa M, Kobayashi S, Sato Y, Maeshima H, Hozumi Y, Lefor A T, Kurihara K, Sata N and Yasuda Y 2014 *Breast Cancer* **21** 394–401

[21] Motomura K, Izumi T, Tateishi S, Tamaki Y, Ito Y, Horinouchi T and Nakanishi K 2016 *Br. J. Surg.* **103** 60–9

[22] Gleich B and Weizenecker J 2005 *Nature* **435** 1241–7

[23] Kratz H, Taupitz M, Ariza de Schellenberger A, Kosch O, Eberbeck D, Wagner S, Trahms L, Hamm B and Schnorr J 2018 *PLoS One* **13** e0190214

[24] Tanaka S, Hirata A, Saito Y, Mizoguchi T, Tamaki T, Sakita I and Monden M 2001 *IEEE Trans. Appl. Supercond.* **11** 665–8

[25] Shiozawa M, Lefor A T, Hozumi Y, Kurihara K, Sata N, Yasuda Y and Kusakabe M 2013 *Breast Cancer* **20** 223–9

[26] Ookubo T, Inoue Y, Kim D, Ohsaki H, Mashiko Y, Kusakabe M and Sekino M 2016 *Electron. Commun. Jpn.* **99** 13–21

[27] Kuwahata A, Chikaki S, Ergin A, Kaneko M, Kusakabe M and Sekino M 2017 *AIP Adv.* **7** 056720

[28] Kaneko M, Ohashi K, Chikaki S, Kuwahata A, Shiozawa M, Kusakabe M and Sekino M 2017 *AIP Adv.* **7** 056713

[29] Pouw J J, Grootendorst M R, Bezooijen R, Klazen C A, De Bruin W I, Klaase J M, Hall-Craggs M A, Douek M and Ten Haken B 2015 *Br. J. Radiol.* **88** 20150634

[30] Anninga B *et al* 2016 *Ann. Surg. Oncol.* **23** 2070–8

[31] Cousins A, Balalis G L, Thompson S K, Forero Morales D, Mohtar A, Wedding A B and Thierry B 2015 *Sci. Rep.* **5** 10842

Chapter 2

Magnetic separation of organelles using magnetic beads

Mari Takahashi and Shinya Maenosono

2.1 Introduction

A cell is composed of different compartments, so-called cellular organelles, as illustrated in figure 2.1. Cellular organelles play critical roles in metabolism, transport, immunity and homeostasis, which are indispensable for the maintenance of life. There are various proteins on/in cellular organelles that cooperate to result in biological actions. However, the proteins present in many organelles have not been completely identified. The identification and characterization of proteins on/in cellular organelles are quite important, not only for cell biology, but also for drug discovery and clinical diagnosis.

Several conventional techniques exist to identify proteins on/in the cellular organelles of interest. Gene manipulation, including gene knockdown and gene knockout, is a powerful tool that is frequently used in the field of biology. If a phenotype is changed after disruption or suppression of a specific gene, the change can be attributed to the protein encoded by the gene. In this manner, the function of the protein can be revealed. Unfortunately, it is not always easy to identify the proteins that are actually active on/in cellular organelles solely by gene manipulation both because a gene and its protein do not always have a one-to-one relationship and gene manipulation cannot provide information about post-translational modifications.

The isolation of target cellular organelles has attracted attention as a complementary technique because it enables the direct analysis of intact proteins on/in the isolated organelles. Several commonly used centrifugation techniques for organelle isolation are briefly described here. To isolate cellular organelles by centrifugation, the cell membrane must first be disrupted by one of the following methods: (i) osmotic shock method, (ii) sonication method, (iii) mild detergent method, (iv) French press method, or (v) tissue grinder method (figure 2.2) [1]. All of these methods possess advantages and disadvantages. For example, physical disruption

doi:10.1088/978-0-7503-1584-5ch2

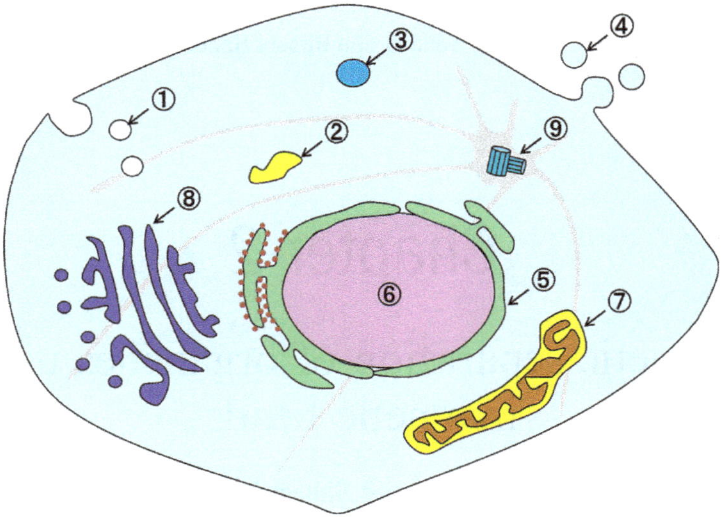

Figure 2.1. Illustration of an animal cell: ① endosome, ② lysosome, ③ peroxisome, ④ exosome, ⑤ endoplasmic reticulum, ⑥ nucleus, ⑦ mitochondria, ⑧ Golgi body and ⑨ centrioles.

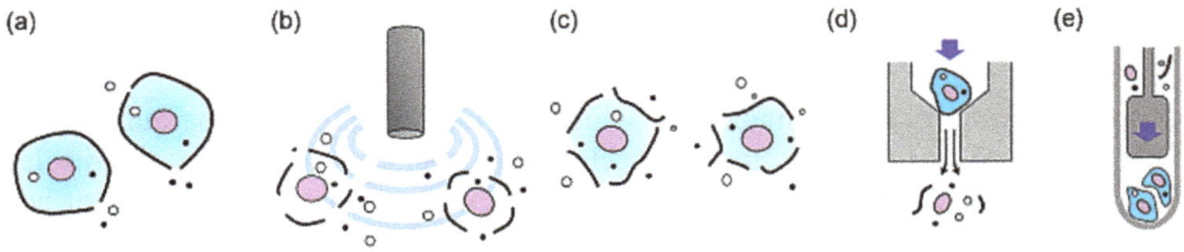

Figure 2.2. Schematic illustrations of several methods for cell membrane disruption: (a) osmotic shock method, (b) sonication method, (c) mild detergent method, (d) French press method and (e) tissue grinder method.

methods often cause deformation and/or aggregation of proteins owing to the heat produced by the equipment and thus need to be performed at low temperature. In addition, these methods are poorly reproducible in terms of cell membrane disruption efficiency. Meanwhile, chemical disruption methods require strict experimental procedures under appropriate conditions; otherwise, the proteins on the cellular organelles can become dissolved. The solution obtained after gentle disruption of the cell membrane, designated as the cell homogenate, contains intact cellular organelles. Differential centrifugation is a widely used technique to isolate certain cellular organelles. First, the homogenate is centrifuged at low speed to separate intact cells, nuclei and cytoskeletons from the other cellular components. Second, the supernatant is centrifuged at medium speed to precipitate mitochondria, lysosomes and peroxisomes. Third, the supernatant is centrifuged at high speed to obtain microsomes and other small vesicles. Finally, the supernatant is centrifuged at very high speed to obtain ribosomes, viruses and large macromolecules, as shown in figure 2.3(a).

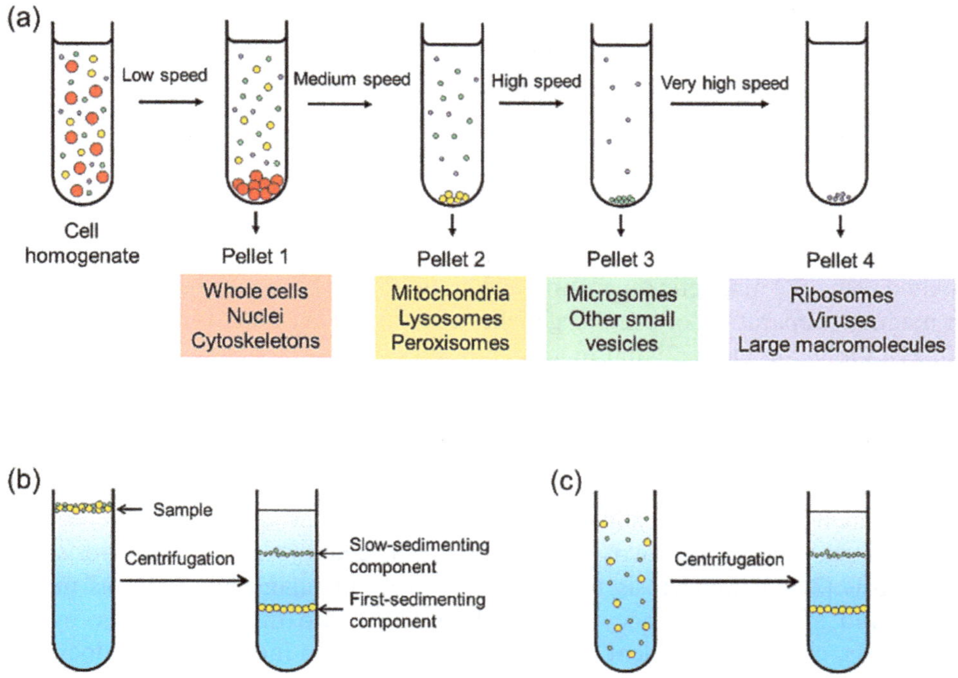

Figure 2.3. Schematic illustrations of several commonly used centrifugation techniques for organelle isolation: (a) differential centrifugation, (b) rate zonal centrifugation and (c) isopycnic centrifugation.

Density gradient centrifugation can be used to separate cellular organelles based on their physical properties such as size, mass and density. There are two typical methods for density gradient centrifugation: rate zonal centrifugation, also called velocity sedimentation (figure 2.3(b)), and isopycnic centrifugation, also called sedimentation equilibrium (figure 2.3(c)) [1]. In the rate zonal centrifugation method, a sample solution is gently layered on a sucrose gradient solution (typically 5–20 wt%), preliminarily prepared in a centrifuge tube, and then subjected to ultracentrifugation. After centrifugation, several bands are formed within the tube depending on the buoyant density of each cellular component. In this way, different organelles of similar size and mass can be separated. The fractionated component is then subjected to column chromatography or gel electrophoresis to purify the proteins present in the fraction. In the isopycnic centrifugation method, a sample solution is mixed with a sucrose or cesium chloride solution and then subjected to ultracentrifugation. During the centrifugation, a density gradient is spontaneously formed in the solution, and thus different organelles of similar size and mass can be separated, similar to the case for the rate zonal centrifugation method. Unfortunately, heterologous organelles that are barely different in size and mass cannot be effectively separated by these centrifugation techniques. Moreover, superficial proteins on the organelles can become detached during the lengthy and harsh centrifugation processes. Therefore, the demand for novel techniques for the isolation of cellular organelles with maintenance of their intracellular state has increased to identify proteins and/or lipids on/in these organelles.

In response to this situation, magnetic separation of cellular organelles has become an important technique in recent years because of its ability to selectively isolate target cellular organelles rapidly and mildly irrespective of size and mass. As described previously, magnetic separation of cells and/or proteins has been widely used for medical diagnosis and therapy. In contrast, magnetic separation of cellular organelles is much less common than magnetic cell/bacteria/protein separation and often used only in the field of basic biology. As described in the following sections, endosomes [2–6], exosomes [7–12] and mitochondria [13–16] have been successfully isolated using the magnetic separation technique. There have also been some reports on magnetic separation of other organelles such as peroxisomes [17–19], lysosomes [20, 21], Golgi vesicles [22] and macropinosomes [23].

2.2 Magnetic separation of endosomes

Endosomes are small membrane vesicles formed by an endocytic process in a cell, during which extracellular materials or receptors on the cell membrane become incorporated. There are two general types of endocytosis: phagocytosis and pinocytosis [24]. Phagocytosis takes place actively in phagocytes such as macrophages and monocytes. Pinocytosis can be divided into three categories: macropinocytosis, clathrin-dependent endocytosis, and clathrin-independent endocytosis. Extracellular materials incorporated through endocytosis are transported to lysosomes, which results in degradation of the materials, or returned to the cellular membrane for recycling of the receptors. During this material transport process, endosomes change their state (early endosomes, late endosomes and recycling endosomes). Early endosomes undergo differentiation into late endosomes or recycling endosomes depending on the situation. In the case of receptor-mediated endocytosis, for example, both ligand and receptor are simultaneously incorporated. The ligand is finally transferred to a lysosome, followed by degradation, while the receptor is returned to the cell membrane and reused several hundred times by recycling endosomes [25]. Endocytosis regulates various cellular processes, including cell motility, cell determination, nutrient uptake and microbial invasion [3]. Therefore, functional disorders of endocytosis can cause many diseases such as familial hypercholesterolemia [26] and cancer [27]. Low-density lipoprotein (LDL) is incorporated into cells through receptor-mediated endocytosis. Patients with familial hypercholesterolemia have dysfunction of the LDL receptor and thus cannot uptake LDL appropriately [26]. For this reason, their blood LDL concentration increases, thereby enhancing the potential for arteriosclerosis, which may lead to serious diseases such as myocardial infarction.

Niemann–Pick type C (NPC) disease belongs to a group of lysosomal storage diseases, and affected patients cannot properly metabolize LDL-derived cholesterol. Most patients with NPC disease (~95%) have mutations in the *NPC1* gene, while the remaining patients have mutations in the *NPC2* gene. Although the role of the NPC2 protein in the pathogenesis of NPC disease is unclear, deficiency in either the *NPC1* gene or the *NPC2* gene produces indistinguishable cellular phenotypes. Based on these observations, it can be hypothesized that a lack of NPC1 protein affects the function and/or location of NPC2 protein. To test this hypothesis, Chen *et al* [2] isolated late endosomes from wild-type (Wt) mice and mice with mutations in NPC1 protein (designated as NPC1 mice) by magnetic separation using dextran-coated superparamagnetic iron oxide nanoparticles (SPIONs), which were synthesized by coprecipitation of $FeCl_2$ and $FeCl_3$ [20]. They intravenously injected the dextran-coated SPIONs into a mouse tail vein. Since SPIONs cannot be metabolized, they accumulate in the liver and are endocytosed by liver cells. The mice were euthanized at 0, 0.5, 1, 2 or 4 h after the SPION injection, and their livers were excised, homogenized and centrifuged. The obtained postmitochondrial supernatant was subjected to magnetic separation. The magnetically separated fraction was analyzed by transmission electron microscopy (TEM) and western blotting (WB). Maximal association of Rab5 was observed between 0.5 and 1 h after the injection, while maximal association of Rab7 was observed at 2 h after the injection. Meanwhile, the α-GalNA case showed maximal association at 4 h after the injection. These findings indicate that specific organelles (early endosomes, late endosomes and lysosomes) can be enriched by controlling the time after injection of SPIONs. They harvested late endosomes from Wt and NPC1 mice by magnetic separation and then analyzed the lipid profiles of the late endosomes by high-performance thin-layer chromatography. Thus, a significant difference in the lipid profiles of late endosomes was observed between the Wt and NPC1 mice.

Another example of magnetic separation of endosomes was reported by Wittrup *et al* [3, 4]. Macromolecular ligands, such as polylysine/cationic lipid–DNA complexes, cationic polymers and antimicrobial peptide–DNA complexes, are internalized by cell-surface heparin sulfate proteoglycan (HSPG)-mediated endocytosis [28]. As some viruses also use HSPG-mediated internalization for entry into host cells, HSPGs have become an important target molecule for drug delivery [29]. However, the exact function of HSPGs in the endocytotic process has remained unclear, such as whether HSPGs are true internalizing receptors or just receptors for initial cell-surface attachment. Wittrup *et al* [3] prepared AO4B08 antibody (HSPG ligand)-conjugated superparamagnetic nanoparticles and added them to HeLa cells together with a fluorophore-conjugated AO4B08 antibody. After incubation for a certain duration, the cells were washed and treated with trypsin. The postnuclear supernatant (PNS) was then harvested by centrifugation. Subsequently, the PNS was subjected to magnetic separation using a magnetic separation tool (PickPen; Bio-Nobile). By monitoring the fluorescence before and after the magnetic isolation, it was possible to roughly estimate the yield and purity of the isolated vesicles. The results indicated that the magnetic isolation successfully enriched the AO4B08 antibody-conjugated SPION-containing endocytic vesicles with high yield. They further determined the HSPG concentration in the magnetically isolated vesicles using metabolically labeled ^{35}S-sulfate. Thus, it was observed that HSPGs were enriched by approximately 20 fold in the magnetically isolated fraction compared with the PNS and nonmagnetic fractions. These results confirmed that HSPGs were internalized into the cell by endocytosis, which indicated that HSPGs are true internalizing receptors for macromolecular cargo [3].

In another study, Li *et al* [5] separated insulin receptor-mediated endosomes. Cell signaling and endocytosis have traditionally been regarded as distinct processes. However, it has recently been recognized that these processes are closely linked, and endocytosis has been found to regulate receptor signaling cascades [30, 31]. Therefore, elucidation of the relationships between signaling and endocytosis has become an important topic in the field of cell biology. The Ras/Raf-1/MEK1/Erk1-2 cascade regulates cell proliferation and differentiation, even though the precise role of this signaling cascade in endocytosis remains controversial [32]. When insulin binds to an insulin receptor expressed on a cell surface, the receptor initiates a phosphorylation cascade [33]. Li *et al* [5] prepared insulin-conjugated ferrofluid nanoparticles (Fe-INS) with a diameter of 10 nm simply by mixing ferrofluid (FerroTec) with insulin molecules. They successfully internalized Fe-INS into HIRcB cells through clathrin-coated vesicles by incubating Fe-INS with HIRcB cells. It was found that the size of the vesicles containing Fe-INS became larger with increasing incubation time from 5 min to 60 min. Specifically, the vesicle size was approximately 100–200 nm after a 5 min incubation but increased to approximately 1–2 μm after a 60 min incubation. The HIRcB cells containing Fe-INS were homogenized and allowed to flow through a magnetic column placed on a magnetized column holder. The magnetically attracted fraction was then eluted from the column after its removal from the holder. The eluted fractions with different incubation times were analyzed by WB, and the results revealed that the localized Fe-INS underwent a transition from early endosomes to lysosomes with increasing incubation time without contamination by endoplasmic reticulum or Golgi. Interestingly, when they used bovine serum albumin-conjugated ferrofluid nanoparticles (Fe-BSA), these nanoparticles were not internalized into the cell. Li *et al* [5] also investigated the activation of proteins involved in the Ras/Raf-1/MEK/Erk cascade in a time-dependent manner. Their results indicated that the signaling cascade takes place on endocytic vesicles that are trafficked, along with insulin and insulin receptors.

2.3 Magnetic separation of exosomes

Cells release different types of extracellular membrane vesicles, such as exosomes (~40–100 nm), microvesicles (~50 nm to several micrometers) and apoptotic bodies (~50–5000 nm) [34]. Exosomes are specific subtypes of extracellular membrane vesicles [35] and have been widely studied at both biochemical and functional levels. Exosomes are secreted upon fusion of multivesicular endosomes with the cell surface. In general, exosomes are classified according to their size, density and expression of certain markers such as CD63, CD81 and CD82 [36]. They have pleiotropic biological roles, including antigen presentation and intercellular transfer of protein cargo, mRNA, microRNA, lipids and oncogenic potential [10]. Exosomes are also involved in cancer, neurodegenerative disease progression, cardiovascular disease and infectious disease [37, 38]. Differential centrifugation and ultracentrifugation are conventional methods for the isolation of exosomes [37]. Several exosome detection kits are commercially available. The main advantages of magnetic separation over the conventional separation techniques are higher accuracy, shorter processing time and separation capability of certain exosomes from other types of exosomes, as described in the following.

Tauro *et al* [9, 10] isolated exosomes by magnetic separation. First, they compared the exosome isolation efficiencies among different separation techniques, including ultracentrifugation, differential centrifugation and immunomagnetic separation [9]. As a result, immunomagnetic separation was found to enrich exosomes more effectively than the other methods. Next, they investigated the physiological role of exosomes in colorectal cancer biology [10]. Because the tumor microenvironment plays critical roles in tumor initiation, progression and metastasis, detailed characterization of exosomes, including proteomics analyses, is important to understand their possible roles in tumor microenvironment regulation. They cultured human colon carcinoma LIM1863 cells to grow organoids. Two distinct populations of exosomes exist in a culture medium of LIM1863 cells: A33-exosomes and EpCAM-exosomes. These two types of exosomes are indistinguishable from one another by TEM observation and both contain general exosome markers such as TSG101, Alix and HSP70. For isolation of the two types of exosomes, a culture medium was centrifuged at low speed to remove the intact cells, cell detritus and large membrane particles. The supernatant was then centrifuged to remove shed microvesicles. Subsequently, the supernatant was filtered and concentrated to obtain the concentrated culture medium (CCM). Anti-A33-coupled magnetic beads composed of SPIONs (Dynabeads; Invitrogen) were mixed with the CCM, followed by magnetic separation. After separation of the A33-exsosomes, the residual CCM was treated with anti-EpCAM-coupled SPION-based magnetic beads (EpCAM MicroBeads; Miltenyi Biotec) to isolate EpCAM-exosomes. Two common exosome markers, Alix and TSG101, were detected in both A33-exosomes and EpCAM-exosomes by WB. To clearly distinguish A33-exosomes from EpCAM-exosomes, a proteomics analysis was performed. Thus, 1024 and 898 proteins were identified in A33-exosomes and EpCAM-exosomes, respectively. Of these, 340 and 214 proteins were found to be unique to A33-exosomes and EpCAM-exosomes, respectively, while the remaining 684 proteins were common to both types of exosomes. Interestingly, several MHC class I molecules were exclusively observed in A33-exosomes, while neither MHC class I nor MHC class II molecules were observed in EpCAM-exosomes. Colocalization of EpCAM, claudin-7 and CD44 was observed in EpCAM-exosomes. These proteins interact with one another to promote tumor progression.

Clayton *et al* [7] proposed a rapid and versatile technique for the analysis of membrane proteins on exosomes based on immunomagnetic separation and flow cytometry. They used five cell lines: B-lymphocytes (B cells) derived from two donors [B-LCL(DA) and B-LCL(MA)], human Burkitt lymphoma line (Daudi) cells deficient in MHC class I, human chronic myelogenous leukemia line (K562) cells deficient in MHC class II and monocyte-derived dendritic cells. It should be noted that the B-lymphocytes were immortalized with the Epstein–Barr virus in advance. The culture medium was centrifuged from low to high speed to remove intact cells and cell debris. Magnetic particles with a mean diameter of 4.5 μm conjugated with anti-human leukocyte antigen antibodies were then added to the cell-free supernatants, followed by magnetic separation. Exosomes were magnetically separated from the B-LCL(DA), B-LCL(MA), Daudi, K562 and dendritic cell supernatants. Figure 2.4 shows TEM images of a single magnetic particle coated with extracellular vesicles derived from B-LCL(DA) cells. Based on a size analysis,

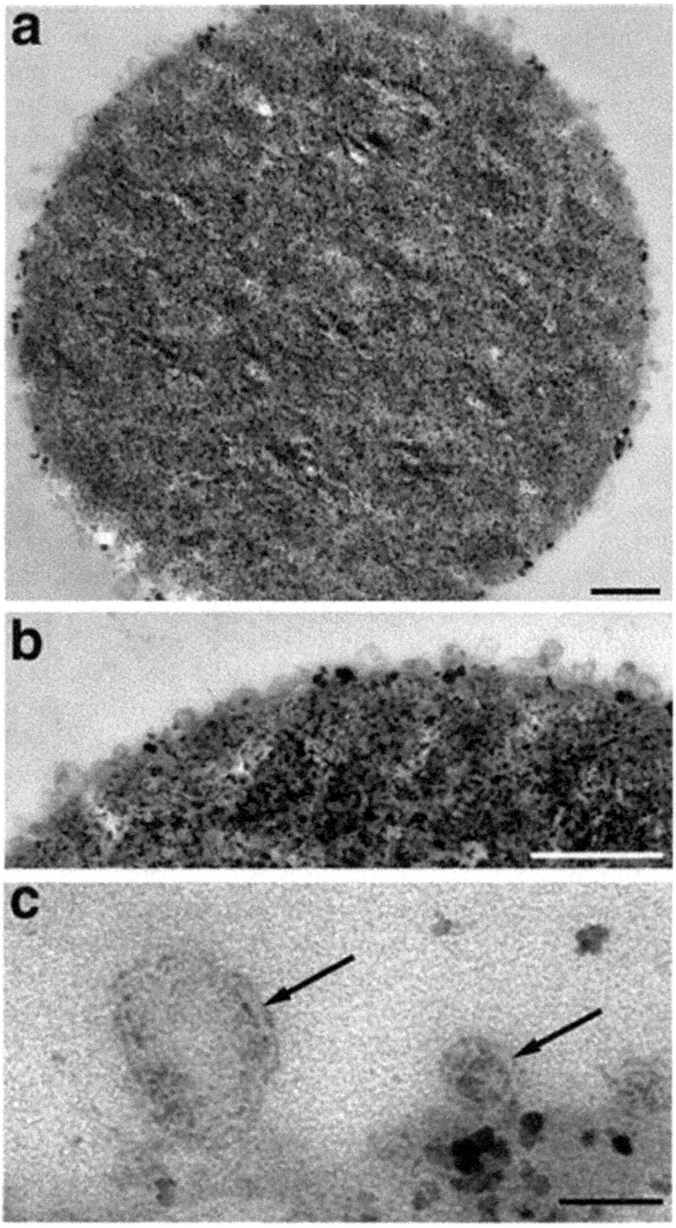

Figure 2.4. Ultrathin sections of exosome-coated beads. Beads were coated under saturating conditions, with exosomes derived from B-LCL(DA) cells and ultrathin sections viewed under a transmission electron microscope. The bead surface is well coated with vesicles: (a) bar = 0.5 μm; (b) bar = 0.5 μm at higher magnification. (c) At a higher magnification, membrane structures on two different sized vesicles are visible (arrows), bar = 50 nm. (Copyright 2001 Elsevier. Reprinted with permission [7].)

~70% of the vesicles were thought to be exosomes. Next, FITC-conjugated IgG antibodies, which were conjugated to magnetic particles, were selectively bound to specific proteins expressed on exosomes for flow cytometric analysis. The results clearly showed that B-cell-derived exosomes expressed MHC class I and MHC class II molecules. Daudi-cell-derived exosomes were most strongly stained for MHC class II molecules and there was no staining for MHC class I molecules, as expected. Meanwhile, the K562 cell line, which is deficient in MHC class II molecules, did not secrete material that became bound to the magnetic particles. The dendritic-cell-derived exosomes expressed abundant MHC class I and II molecules. The merit of their technique is that it requires a short time (approximately half a day) for qualitative and semiquantitative analysis of proteins on exosomes [7].

Zong *et al* [12] combined immunomagnetic separation and surface-enhanced Raman scattering (SERS)-based sensing techniques to detect tumor-cell-derived exosomes with enhanced sensitivity. Tumor-cell-derived exosomes are heavily involved in the modulation of the tumor microenvironment. Therefore, tumor-cell-derived exosomes are considered to be important cancer biomarkers, and analysis of these exosomes can provide insightful information on the pathological stages of cancerous diseases. Zong *et al* synthesized two nanoparticles: Fe_3O_4@SiO_2 core@shell nanoparticles (size of approximately 120–220 nm) and Au@Ag@SiO_2 core@shell@shell nanorods containing Raman reporter molecules in their structure. Anti-CD63 and anti-HER2 antibodies were conjugated to the surface of the Fe_3O_4@SiO_2 nanoparticles (magnetic probes) and Au@Ag@SiO_2 nanorods (SERS probes), respectively. In their study, SKBR-3 and normal human embryonic lung fibroblasts (MRC-5) were used as model cells for cancerous and normal cells, respectively. An exosome stock solution was prepared from a cell culture medium using an exosome precipitation kit (Exo-Quick-TC; System Biosciences). Next, the SERS and magnetic probes were simultaneously added to the exosome stock solution to form sandwich-type immunocomplexes (i.e. Fe_3O_4@SiO_2/exosome/Au@Ag@SiO_2). After incubation, the immunocomplexes were collected with a magnet and subjected to SERS measurements. Based on a quantitative analysis, the SERS-based method showed high sensitivity, with a detection limit of 1200 exosomes. The SERS signal of SKBR-3 cell-derived exosomes was 3.8 times higher than that of MRC-5 cell-derived exosomes. These findings are consistent with the fact that tumor cells usually secrete more exosomes than normal cells [39]. This SERS-based method requires only 2 h from collection to identification of the exosomes, being much quicker than conventional methods that normally require approximately 12 h in total [12].

2.4 Magnetic separation of mitochondria

Mitochondria participate in various biological processes, including energy generation, production of oxygen free radicals, maintenance of intracellular calcium homeostasis, signal transduction and apoptosis [14]. Therefore, mitochondrial dysfunction often leads to the development of various diseases [40]. It has been reported that mitochondrial dysfunction is related to obesity [41], cancer [42], neurodegenerative disease [43] and cardiac disease [44], among others. For a precise investigation of the role and performance of functional mitochondria, the isolation of mitochondria is crucially important, and thus, several methods, such as density gradient centrifugation and differential centrifugation, have been used towards this purpose. Although differential centrifugation, which was developed during the 1950s [45] and subsequently modified in various ways, is the most frequently used method, it has limitations, such as being a time-consuming process and including a lack of purity. Meanwhile, density gradient centrifugation can isolate mitochondria with high purity but also requires a long time and shows contamination depending on the sample type [15]. Furthermore, the bilayer structure of mitochondria can be destroyed under certain physiological conditions. As these conventional separation techniques have problems in terms of purity, reproducibility and ease of process, many researchers have become interested in magnetic isolation of mitochondria.

Franko *et al* [15] proposed a standardized isolation method for mitochondria by demonstrating their isolation from various mouse tissues. They used anti-TOM22 (translocase of outer mitochondria membrane 22 homolog) antibody-coupled magnetic beads (MicroBeads; Miltenyi Biotec) for the magnetic separation of mitochondria. To magnetically separate mitochondria, tissue lysates were prepared using a Mitochondria Extraction Kit (Miltenyi Biotec). The tissue lysates were treated with the anti-TOM22 MicroBeads, followed by magnetic separation using a MACS separator (Miltenyi Biotec). After the magnetic separation, the magnetically labeled eluted fraction and nonlabeled (wash) fraction were analyzed by TEM and WB. Figure 2.5 shows TEM images of magnetically isolated mitochondria from the heart (figure 2.5(a)), muscle (figure 2.5(b)), brain (figure 2.5(c)) and liver (figure 2.5(d)). The small black dots observed outside the membrane in these images correspond to the anti-TOM22 MicroBeads. By carefully examining the morphology of the isolated mitochondria, it was revealed that ~90% of the mitochondria were intact. WB analysis was performed to detect mitochondrial markers (TOM20 and cytochrome c) and confirm the presence of mitochondrial proteins in the eluted fractions. Thus, both TOM20 and cytochrome c were detected in the eluted fractions from the heart, brain and liver tissues, while none of these markers were detected in the wash fractions. In contrast, PDIA2 (endoplasmic reticulum marker) and PEX1 (peroxisomal protein marker) were rarely detected in the eluted fractions, while they were detected in the wash fractions. These results suggested that mitochondria were successfully enriched. To assess the function of the magnetically separated mitochondria, the oxygen consumption rates of mitochondria derived from the skeletal muscle, liver and brain were analyzed by the Clark electrode method. The results revealed that the magnetically separated mitochondria from these tissues maintained their function.

Finally, Franko *et al* [15] then compared their magnetic separation technique with differential centrifugation in terms of purity and quality of the isolated mitochondria. In the case of differential centrifugation, both TOM20 and cytochrome c were detected in the precipitate. Unfortunately, PDIA2 was also detected in the precipitate. These results suggested that the separation of mitochondria and endoplasmic reticulum was not successful using differential centrifugation. They

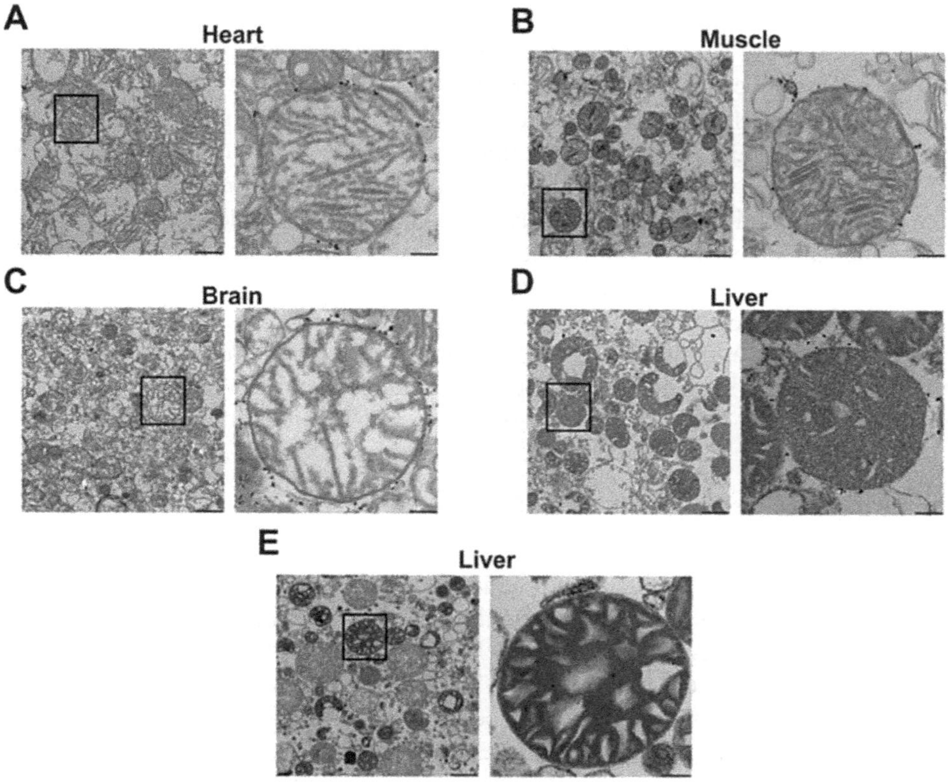

Figure 2.5. The anti-TOM22 magnetic beads method yields intact mitochondria. The anti-TOM22 magnetic bead method was applied to purify mitochondria from (a) heart, (b) skeletal muscle, (c) brain and (d) liver tissues. Mitochondria were isolated by a differential centrifugation technique from (e) liver tissue. Mitochondrial morphology was investigated by TEM from four biological samples per group and representative pictures are shown. Pictures were taken at 4000× (left panel) and 20 000× (right panel) magnification and black bars display 1 mm (left panel) and 200 nm (right panel). Black dots at the outer membrane in a, b, c and d represent anti-TOM22 magnetic beads. (Copyright 2013 Franko A *et al.* Reprinted with permission [15].)

further confirmed that the purity of their magnetically separated mitochondria was higher than that of mitochondria separated by differential centrifugation using liquid chromatography–tandem mass spectrometry.

Tang *et al* [14] also compared a magnetic separation technique for mitochondria with other conventional separation methods such as differential centrifugation and density gradient centrifugation. They conjugated an anti-monoamine oxidase-A (MAO-A; mitochondrial marker) antibody to Fe_3O_4@SiO_2 core@shell nanoparticles with a mean diameter of approximately 220 nm. Meanwhile, HepG2 and HeLa cells were cultured and then homogenized. Anti-MAO-A-conjugated Fe_3O_4@SiO_2 nanoparticles were added to the cell lysates, followed by incubation for 20 min at 4 °C. Subsequently, a magnet was placed on the tube wall to collect the magnetic nanoparticles. The supernatant was discarded and the magnetically attracted substances on the tube wall were washed. After washing, the magnetically attracted fraction was collected. The authors revealed that the purity of their isolated mitochondria was more than 90%, while the mitochondria isolated by differential centrifugation were contaminated by other subcellular components. In addition, in the case of HepG2 cells, the magnetic separation method gave a twofold higher yield than differential centrifugation and a threefold higher yield than density gradient centrifugation. The respiratory functions of the mitochondria isolated by the different methods were compared, and the magnetically isolated mitochondria were found to exhibit the highest activity. It was also confirmed that the purity of the magnetically separated mitochondria was higher than that of the mitochondria separated by either differential centrifugation or density gradient centrifugation. These results showed that magnetic separation was a superior isolation technique with higher yield and greater mitochondrial purity and activity compared with conventional centrifugation techniques. In addition, the magnetic separation requires only about 1 h from cell harvesting to isolation of mitochondria.

2.5 Magnetic separation of autophagosomes

Autophagy is an intracellular bulk degradation system that is conserved among eukaryotes from yeast to humans. It is induced by cellular stresses, such as nutrient starvation, and generates energy and building blocks for biosynthesis through the degradation of cytoplasmic components. Dysfunction of autophagy leads to various diseases, such as cancer, metabolic dysfunction and vascular instability, and in organ-specific pathologies, such as neurodegeneration, cardiomyopathies and myopathies, non-alcoholic fatty liver disease or Crohn's disease [46]. In general, it is known that there are three types of autophagy: macroautophagy, microautophagy and chaperon-mediated autophagy. In macroautophagy, cytoplasmic constituents are first enwrapped with an isolation membrane followed by closure of the membrane resulting in the formation of autophagosomes. Eventually, the autophagosomes fuse with lysosomes to form autolysosomes and the contents of the autophagosomes are degraded. However, autophagy can also selectively target intracellular pathogens. This type of autophagy is so-called xenophagy and plays a critical role in controlling bacterial infection [47].

Takahashi *et al* [48] developed ultrasmall magnetic nanoparticles that have both imaging and magnetic separation capabilities. In particular, monodispersed Ag@FeCo@Ag core@shell@shell magnetic-plasmonic hybrid nanoparticles with a mean diameter of 15 nm were chemically synthesized (figure 2.6). The Ag core and FeCo shell provide the imaging and magnetic separation capabilities, respectively. Because the surfaces of the as-synthesized hybrid nanoparticles are covered with hydrophobic organic ligands, the surfaces of the nanoparticles were then modified with thiolated ε-poly-L-lysine (PLL-SH). PLL-SH-modified hybrid nanoparticles were transfected into COS-1 cells by lipofection. At an appropriate time after lipofection, the cells were fixed, stained for Vps26 (an early endosomal marker protein) or LC3 (an autophagosomal marker protein), and examined by a confocal laser scanning microscope (CLSM). Thirty minutes after transfection, the hybrid nanoparticles partly co-localized with Vps26 (figure 2.7(A)), but not with LC3 (figure 2.7(C)). One hour after lipofection, a few nanoparticle-positive structures

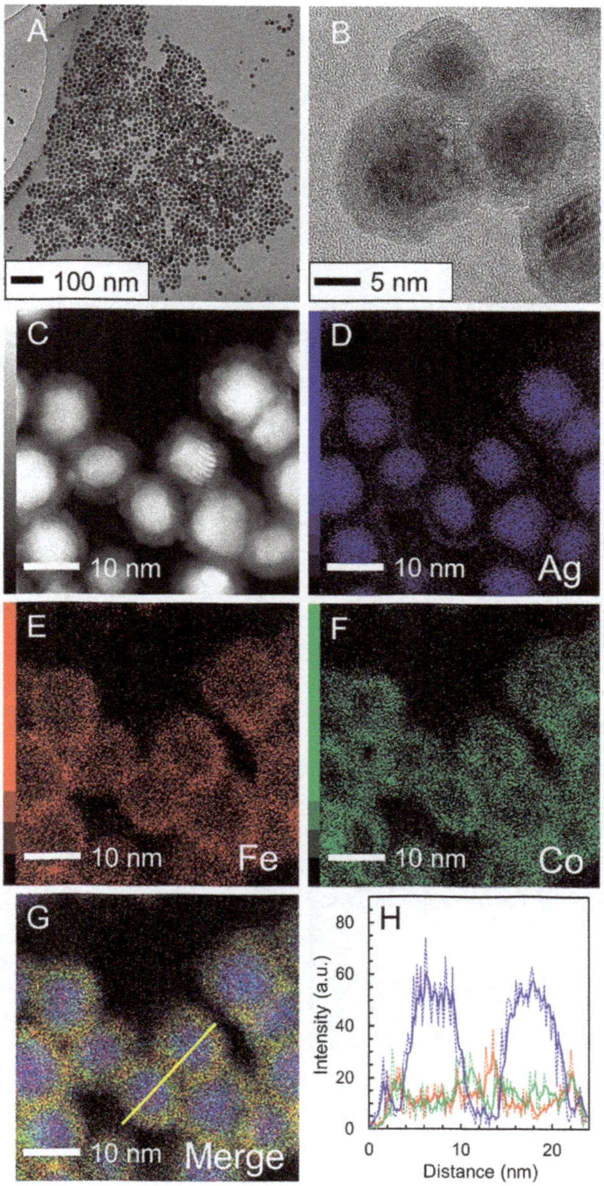

Figure 2.6. Morphology of the hybrid nanoparticles. (A) TEM, (B) HR-TEM, and (C) STEM-HAADF images of Ag@FeCo@Ag core@shell@shell nanoparticles. (D) Ag L edge, (E) Fe K edge, (F) Co K edge, and (G) merged images of EDS elemental mapping images for Ag@FeCo@Ag nanoparticles. (H) EDS line profile along the yellow line in (G). The dashed and solid lines are the raw and low-pass-filtered profiles, respectively. The blue, green and red colors represent Ag L-, Co K- and Fe K-edge intensities, respectively. (Copyright 2017 American Chemical Society. Reprinted with permission [48].)

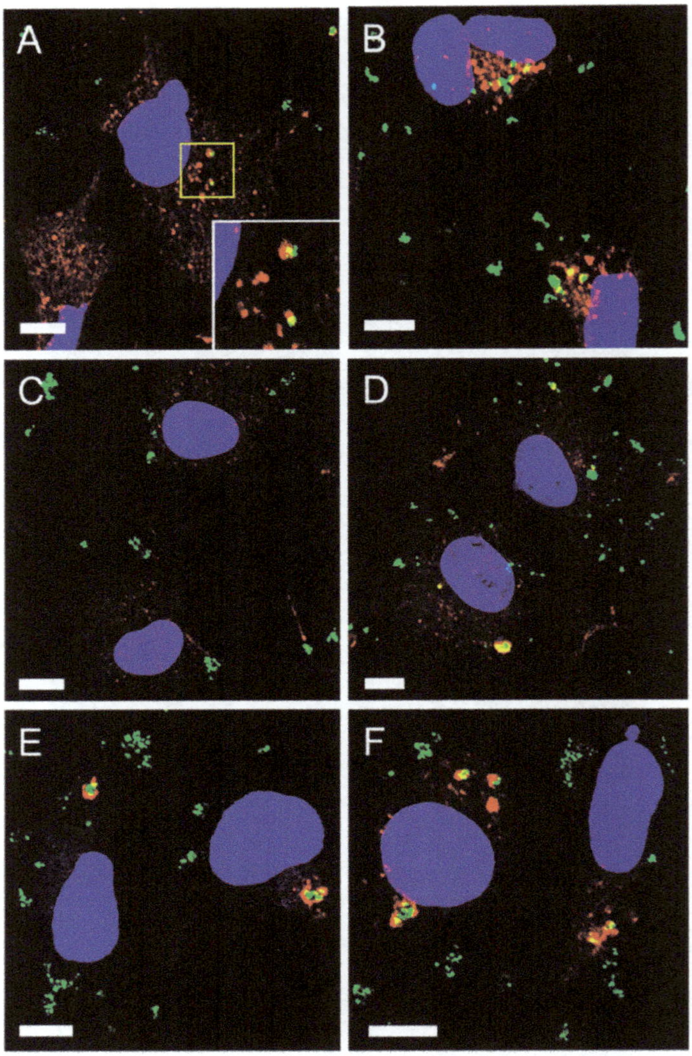

Figure 2.7. CLSM images of Ag@FeCo@Ag nanoparticle-transfected COS-1 cells. After incubation for (A) 30 min and (B) 1 h for Vps26 staining, and (C) 30 min, (D) 1 h, (E) 2 h, and (F) 4 h for LC3 staining. Blue and green colors represent DAPI-stained nuclei and Ag@FeCo@Ag nanoparticles, respectively. The red color represents Vps26 (A,B) or LC3 (C–F). All scale bars correspond to 10 μm. The inset shown at the bottom right of panel (A) is a magnified view of the region enclosed by the yellow line in panel (A). (Copyright 2017 American Chemical Society. Reprinted with permission [48].)

emerged that were surrounded by LC3 (figure 2.7(D)). These structures were mostly evident 2 and 4 h after lipofection (figures 2.7(E) and (F)). These results indicated that the nanoparticles were first delivered into early endosomes, and then targeted to autophagosomes. It should be noted that the imaging of hybrid nanoparticles was achieved by using plasmon scattering (not fluorescence).

The cells lipofected with hybrid nanoparticles were detached from the culture dish using a cell lifter at the appropriate time and homogenized. The resultant cell lysates were subjected to magnetic separation using an autoMACS Pro Separator (Miltenyi Biotec, Bergisch Gladbach, Germany). The magnetically separated fraction was then eluted from the column, briefly spun, and resuspended in phosphate-buffered saline (PBS). The cell lysate and magnetically separated fractions (MS) were subjected to SDS-PAGE and blotted for LC3, transferrin receptor (TfnR, an endosomal protein), LAMP2 (lysosomal protein), and GAPDH (cytosolic protein). As shown in figure 2.8(B), at time 0, the proteins of interest were not detected in the MS. After 15 min, LC3-II (a membrane-bound form of LC3), TfnR and LAMP2, but not GAPDH, were detected in the MS. The level of LC3-II and LAMP2 in the MS increased up to 8 h, whereas that of TfnR plateaued after 2 h. Throughout the

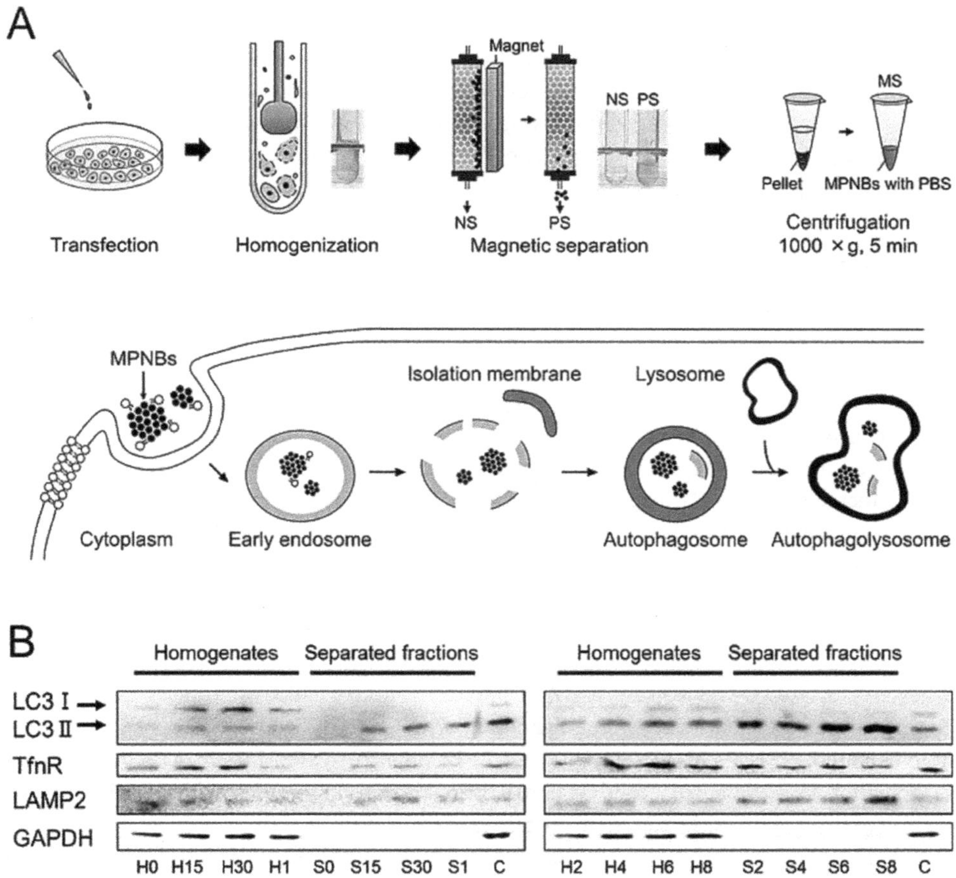

Figure 2.8. Magnetic separation scheme and the results of western blotting. (A) Schematic illustrations of (top) the magnetic separation protocol and (bottom) the change in the localization of hybrid nanoparticles in a cell. (B) The incubation time dependence of the presence of marker proteins in the homogenate (designated as H) and the separated fractions (MS, designated as S). C is a control in which the transfection reagent was present but nanoparticles were absent. The number added after the initial of the sample (H or S) denotes the incubation time, that is, 0, 15, and 30 correspond to 0, 15, and 30 min incubations, respectively, while 1, 2, 4, 6, and 8 correspond to 1, 2, 4, 6, and 8 h incubations, respectively. (Copyright 2017 American Chemical Society. Reprinted with permission [48].)

separation experiments, GAPDH and LC3-I (a cytosolic form of LC3) were not detected in the MS. These results indicated that at early time points ($0.5 < t < 2$ h, where t denotes the incubation time) autophagosomes were mainly isolated and, in contrast, at a later time point ($t > 2$ h), autophagolysosomes were isolated.

2.6 Conclusions and future outlook

The magnetic separation technique has recently been applied not only for the isolation of cells, bacteria and proteins, but also for the isolation of cellular organelles to characterize the proteins on/in the target organelles. In this chapter, several examples of magnetic separation of cellular organelles, such as endosomes, exosomes and mitochondria, were reviewed. It has been clearly demonstrated that the magnetic separation technique has advantages over conventional separation techniques, such as differential centrifugation and density gradient centrifugation, in terms of purity and ease of use. By enabling the on-demand isolation of various cellular organelles, developing and using multifunctional probes to identify proteins on/in the organelles and analyzing the function of these proteins, it will become possible to offer new insights not only in basic biology but also in clinical medicine.

References

[1] Alberts B, Bray D, Hopkin K, Johnson A, Lewis J, Raff M, Roberts K and Walter P 2004 *Essential Cell Biology* 2nd edn (London: Garland Science Taylor and Francis)

[2] Chen F W, Gordon R E and Ioannou Y A 2005 NPC1 late endosomes contain elevated levels of non-esterified ('free') fatty acids and an abnormally glycosylated form of the NPC2 protein *Biochem. J.* **390** 549–61

[3] Wittrup A, Zhang S-H, ten Dam G B, van Kuppevelt T H, Bengtson P, Johansson M, Welch J, Moergelin M and Belting M 2009 ScFv antibody-induced translocation of cell-surface heparan sulfate proteoglycan to endocytic vesicles *J. Biol. Chem.* **284** 32959–67

[4] Wittrup A, Zhang S H, Svensson K J, Kucharzewska P, Johansson M C, Moergelin M and Belting M 2010 Magnetic nanoparticle-based isolation of endocytic vesicles reveals a role of the heat shock protein GRP75 in macromolecular delivery *Proc. Natl. Acad. Sci. U.S.A.* **107** 13342–7

[5] Li H-S, Stolz D B and Romero G 2005 Characterization of endocytic vesicles using magnetic microbeads coated with signalling ligands *Traffic* **6** 324–34

[6] Nakamura N, Lill J R, Phung Q, Jiang Z, Bakalarski C, de Maziere A, Klumperman J, Schlatter M, Delamarre L and Mellman I 2014 Endosomes are specialized platforms for bacterial sensing and NOD2 signalling *Nature* **509** 240–4

[7] Clayton A, Court J, Navabi H, Adams M, Mason M D, Hobot J A, Newman G R and Jasani B 2001 Analysis of antigen presenting cell derived exosomes, based on immuno-magnetic isolation and flow cytometry *J. Immunol. Methods* **247** 163–74

[8] Ji H, Chen M, Greening D W, He W, Rai A, Zhang W and Simpson R J 2014 Deep sequencing of RNA from three different extracellular vesicle (EV) subtypes released from the human LIM1863 colon cancer cell line uncovers distinct mirna-enrichment signatures *PLoS One* **9** e110314

[9] Tauro B J, Greening D W, Mathias R A, Ji H, Mathivanan S, Scott A M and Simpson R J 2012 Comparison of ultracentrifugation, density gradient separation, and immunoaffinity

capture methods for isolating human colon cancer cell line LIM1863-derived exosomes *Methods* **56** 293–304
[10] Tauro B J, Greening D W, Mathias R A, Mathivanan S, Ji H and Simpson R J 2013 Two distinct populations of exosomes are released from LIM1863 colon carcinoma cell-derived organoids *Mol. Cell. Proteomics* **12** 587–98
[11] Hong C S, Muller L, Boyiadzis M and Whiteside T L 2014 Isolation and characterization of CD34+ blast-derived exosomes in acute myeloid leukemia *PLoS One* **9** e103310
[12] Zong S, Wang L, Chen C, Lu J, Zhu D, Zhang Y, Wang Z and Cui Y 2016 Facile detection of tumor-derived exosomes using magnetic nanobeads and SERS nanoprobes *Anal. Methods* **8** 5001–8
[13] Hornig-Do H-T, Günther G, Bust M, Lehnartz P, Bosio A and Wiesner R J 2009 Isolation of functional pure mitochondria by superparamagnetic microbeads *Anal. Biochem.* **389** 1–5
[14] Tang B, Zhao L, Liang R, Zhang Y and Wang L 2012 Magnetic nanoparticles: an improved method for mitochondrial isolation *Mol. Med. Rep.* **5** 1271–6
[15] Franko A, Baris O R and Bergschneider E *et al* 2013 Efficient isolation of pure and functional mitochondria from mouse tissues using automated tissue disruption and enrichment with anti-TOM22 magnetic beads *PLoS One* **8** e82392
[16] Kappler L, Li J, Häring H-U, Weigert C, Lehmann R, Xu G and Hoene M 2016 Purity matters: a workflow for the valid high-resolution lipid profiling of mitochondria from cell culture samples *Sci. Rep.* **6** 21107
[17] Lüers G H, Hartig R, Mohr H, Hausmann M, Fahimi H D, Cremer C and Voelkl A 1998 Immuno-isolation of highly purified peroxisomes using magnetic beads and continuous immunomagnetic sorting *Electrophoresis* **19** 1205–10
[18] Wang Y, Taylor T H and Arriaga E A 2012 Analysis of the bioactivity of magnetically immunoisolated peroxisomes *Anal. Bioanal. Chem.* **402** 41–9
[19] Kikuchi M, Hatano N, Yokota S, Shimozawa N, Imanaka T and Taniguchi H 2004 Proteomic analysis of rat liver peroxisome: presence of peroxisome-specific isozyme of Lon protease *J. Biol. Chem.* **279** 421–8
[20] Rodriguez-Paris J M, Nolta K V and Steck T L 1993 Characterization of lysosomes isolated from *Dictyostelium discoideum* by magnetic fractionation *J. Biol. Chem.* **268** 9110–6
[21] Diettrich O, Mills K, Johnson A W, Hasilik A and Winchester B G 1998 Application of magnetic chromatography to the isolation of lysosomes from fibroblasts of patients with lysosomal storage disorders *FEBS Lett.* **441** 369–72
[22] Mura C V, Becker M I, Orellana A and Wolff D 2002 Immunopurification of Golgi vesicles by magnetic sorting *J. Immunol. Methods* **260** 263–71
[23] Journet A, Klein G, Brugière S, Vandenbrouck Y, Chapel A, Kieffer S, Bruley C, Masselon C and Aubry L 2012 Investigating the macropinocytic proteome of *Dictyostelium* amoebae by high-resolution mass spectrometry *Proteomics* **12** 241–5
[24] Conner S D and Schmid S L 2003 Regulated portals of entry into the cell *Nature* **422** 37–44
[25] Maxfield F R and McGraw T E 2004 Endocytic recycling *Nat. Rev. Mol. Cell Biol.* **5** 121–32
[26] Garcia C K, Wilund K and Arca M *et al* 2001 Autosomal recessive hypercholesterolemia caused by mutations in a putative LDL receptor adaptor protein *Science* **292** 1394–8
[27] Mellman I and Yarden Y 2013 Endocytosis and cancer *Cold Spring Harb. Perspect. Biol.* **5** a016949
[28] Sandgren S, Cheng F and Belting M 2002 Nuclear targeting of macromolecular polyanions by an HIV-Tat derived peptide: role for cell-surface proteoglycans *J. Biol. Chem.* **277** 38877–83

[29] Spillmann D 2001 Heparan sulfate: anchor for viral intruders? *Biochimie* **83** 811–7
[30] Platta H W and Stenmark H 2011 Endocytosis and signaling *Curr. Opin. Cell Biol.* **23** 393–403
[31] Di Fiore P P and De Camilli P 2001 Endocytosis and signaling: an inseparable partnership *Cell* **106** 1–4
[32] Andresen B T, Rizzo M A, Shome K and Romero G 2002 The role of phosphatidic acid in the regulation of the Ras/MEK/Erk signaling cascade *FEBS Lett.* **531** 65–8
[33] Boucher J, Kleinridders A and Kahn C R 2014 Insulin receptor signaling in normal and insulin-resistant states *Cold Spring Harb. Perspect. Biol.* **6** a009191
[34] Van der Meel R, Durka M K, Van Solinge W W and Schiffelers R M 2014 Toward routine detection of extracellular vesicles in clinical samples *Int. J. Lab. Hematol.* **36** 244–53
[35] Mathivanan S, Ji H and Simpson R J 2010 Exosomes: extracellular organelles important in intercellular communication *J. Proteomics* **73** 1907–20
[36] Simons M and Raposo G 2009 Exosomes—vesicular carriers for intercellular communication *Curr. Opin. Cell Biol.* **21** 575–81
[37] De Toro J, Herschlik L, Waldner C and Mongini C 2015 Emerging roles of exosomes in normal and pathological conditions: new insights for diagnosis and therapeutic applications *Front. Immunol.* **6** 203
[38] Kalani A, Tyagi A and Tyagi N 2014 Exosomes: mediators of neurodegeneration, neuroprotection and therapeutics *Mol. Neurobiol.* **49** 590–600
[39] Antonyak M A and Cerione R A 2015 Emerging picture of the distinct traits and functions of microvesicles and exosomes *Proc. Natl. Acad. Sci. U.S.A.* **112** 3589–90
[40] Vafai S B and Mootha V K 2012 Mitochondrial disorders as windows into an ancient organelle *Nature* **491** 374–83
[41] Bournat J C and Brown C W 2010 Mitochondrial dysfunction in obesity *Curr. Opin. Endocrinol. Diabetes Obes.* **17** 446–52
[42] Kiebish M A, Han X, Cheng H, Chuang J H and Seyfried T N 2008 Cardiolipin and electron transport chain abnormalities in mouse brain tumor mitochondria: lipidomic evidence supporting the Warburg theory of cancer *J. Lipid Res.* **49** 2545–56
[43] Johri A and Beal M F 2012 Mitochondrial dysfunction in neurodegenerative diseases *J. Pharmacol. Exp. Ther.* **342** 619–30
[44] Wallace K B 2007 Adriamycin-induced interference with cardiac mitochondrial calcium homeostasis *Cardiovasc. Toxicol.* **7** 101–7
[45] De Duve C, Pressman B C, Gianetto R, Wattiaux R and Appelmans F 1955 Tissue fractionation studies. 6. Intracellular distribution patterns of enzymes in rat-liver tissue *Biochem. J.* **60** 604–17
[46] Schneider J L and Cuervo A M 2014 Autophagy and human disease: emerging themes *Curr. Opin. Genet. Dev.* **26** 16–23
[47] Shibutani S T and Yoshimori T 2014 Autophagosome formation in response to intracellular bacterial invasion *Cell. Microbiol.* **16** 1619–26
[48] Takahashi M, Mohan P, Mukai K, Takeda Y, Matsumoto T, Matsumura K, Takakura M, Arai H, Taguchi T and Maenosono S 2017 Magnetic separation of autophagosomes from mammalian cells using magnetic-plasmonic hybrid nanobeads *ACS Omega* **2** 4929–37

Chapter 3

Development of a new affinity nanobead technology and target isolation of bioactive compounds

Takumi Ito, Mamoru Hatakeyama and Hiroshi Handa

3.1 Introduction

Nucleic acids and proteins taking the minute reaction of the organism greatly affect biological phenomena in living bodies from root and trunk. To elucidate these phenomena and mechanisms at the molecular level, it is essential to purify the molecules involved. However, it is extremely difficult to perform such isolations while maintaining the biological activity of the target molecules, which are often destabilized by changes in temperature or pH. In addition, activity is sometimes lost because these molecules are optimized to function under specific environments. Moreover, because many molecules have similar physicochemical properties, very specific purification technologies are required for this purpose. Therefore, insoluble macromolecules, which possess a wide variety of properties capable of overcoming these problems, are used as materials for separation and purification.

For the separation of protein from living cells (i.e. bioseparation), the most common methods for the purification of molecules without diminishing their original activity involve a wide variety of columns, including ion exchange and gel filtration chromatography from crude cell extracts, followed by analysis of the resultant proteins. However, the purity that can be obtained using such methods is sometimes low. Moreover, the techniques involved are complicated and technically challenging. Consequently, purification of a desired protein in large quantities requires a great deal of effort and long periods of time.

To address these problems, affinity purification technology was developed. This approach, which exploits biologically specific binding (bioaffinity), is also called affinity chromatography. Intermolecular forces and aspects of the tertiary structure, including electrostatic attraction, hydrogen bonding, hydrophobic interaction, and van der Waals forces, contribute to the specific affinities between biological

doi:10.1088/978-0-7503-1584-5ch3

materials: e.g. antigen–antibody, enzyme–substrate, enzyme–repressor, enzyme–complement, hormone–receptor, DNA–DNA, DNA–RNA, and DNA–protein interactions. In affinity chromatography, which takes advantage of such specificity, substances (ligands) that specifically interact (via affinity) with the targeted biological molecules are immobilized on an insoluble carrier, and the target molecules are isolated based on their preferential adsorption in comparison with other molecules.

The properties of a suitable carrier include insolubility, chemical and physical stability, capacity for ligand immobilization, minimal non-specific adsorption and low cost. In the earliest attempt at affinity separation, in 1910, Starkenstein *et al* separated amylase based on its interaction with starch [1]. Studies of ligand immobilization methods and non-specific adsorption have been conducted over the intervening years. In 1986, DNA affinity chromatography was reported by Tjian *et al* [2]. This method involves DNA containing repeats of a specific sequence to which a transcription factor (or other DNA-binding protein) can bind. The DNA is immobilized on an agarose gel, which serves as a carrier, using CNBr. After the DNA-immobilized agarose gel and crude cell extract are mixed, purification is performed by the column method. There is no doubt that this method is very useful judging from the cloning performed of a gene from a result of the primary structure analysis of the amino acid of the purified matter; several kinds of protein have been purified by this method [3].

However, it is very difficult to use this method to obtain the target protein at high purity, largely due to adverse effects such as heterogeneous diffusion, retention of the protein in the gel column, the time required for other methods indispensable for increasing purity, and low protein concentration in the yield. Therefore, in collaboration with Professor Haruma Kawaguchi, Faculty of Science and Technology, KEIO University, we designed non-porous beads of submicron size with large relative surface areas [4, 5]. Over the years, we have immobilized not only DNA, but also drug molecules that do not occur naturally in living cells, on the surface of these beads, and then used them to purify and identify bound proteins. In this chapter, we describe a number of notable results obtained using these affinity purification approaches. In addition, we developed polymer beads containing strongly magnetic nanoferrite, making the operation of affinity purification simpler and easier. Below, we also describe affinity purification using these magnetic beads.

3.2 Preparation of SG beads

DNA-immobilized agarose gel beads have two main problems: non-specific protein adsorption and low purification efficiency. To overcome these issues, we needed to develop non-porous, physically stable beads with submicron size, relatively large surface-to-volume ratio, and functional groups that enable easy surface modification. In addition, to limit non-specific protein adsorption, it was necessary to test materials that are moderately hydrophilic.

Among the materials examined, polyglycidyl methacrylate (GMA) was the most suitable, and polystyrene (Poly-St)-GMA copolymer beads satisfied the conditions described above. The Poly-St of the core was non-porous and had the necessary characteristics of hardness, ability to form true spheres, and high dispersibility. The beads were prepared by soap-free emulsion polymerization, using 2,2′-azo bis (2-amidino propane) hydrochloride (V-50) as the starting reagent and divinylbenzene (DVB) as the crosslinker, under a nitrogen atmosphere [6]. However, these beads still exhibited non-specific protein adsorption, and the purification efficiency of the target molecules was very low, largely because the hydrophobic groups derived from Poly-St prevented the bead surface from being moderately hydrophilic. To eliminate these defects, additional GMA was added 2 h after polymerization of the poly (St-GMA) beads, yielding poly [(St-GMA)-GMA] copolymer beads (SG beads) [7]. As shown in figure 3.1, these beads did not exhibit non-specific protein adsorption, and they satisfied all of the requirements for affinity beads described above [8].

SG beads were prepared by the following procedures. St, GMA, DVB, and H2O (1.2, 1.8, 0.04, and 110 g, respectively) were added to a 200 ml four-neck flask, and a stirring seal, stirring stick, Liebig condenser, and Ceram rubber were fitted. The mixture was stirred at 70 °C under a nitrogen atmosphere by a three one motor under agitation (200 rpm) for 30 min. Thereafter, a solution of V-50 in H2O (0.06 and 10 g, respectively) was added to the reaction mixture via the Ceram rubber while the temperature was maintained. After 2 h, 0.3 g of GMA was added, and the mixture was polymerized for a further 22 h. After recovery of the solution, centrifugal separation, decantation, and re-dispersion were repeated three times to remove impurities such as non-reacted monomers. The resultant beads were measured by transmission electron microscopy (TEM) and dynamic light scattering (DLS). Figure 3.2 shows the TEM views of the resultant latex beads. DLS revealed

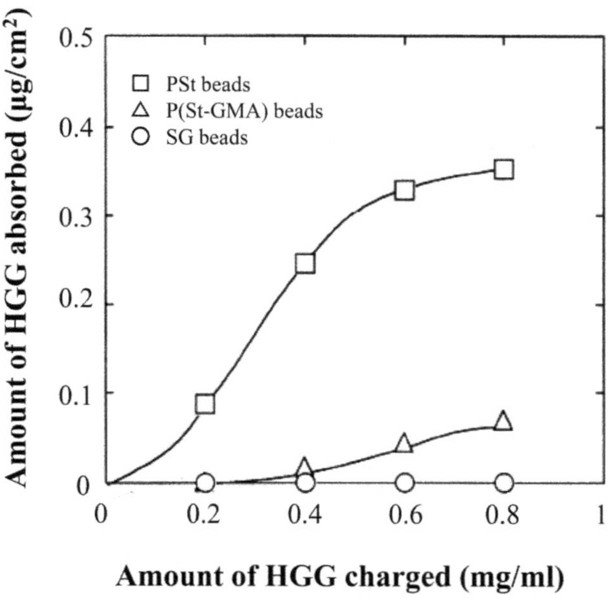

Figure 3.1. Adsorption of human γ-globulins (HGG) on various polymer beads. Square (□),triangle (△), and circle (○) indicate polystyrene, poly (styrene-glycidylmethacrylate [St-GMA]) and SG beads, respectively.

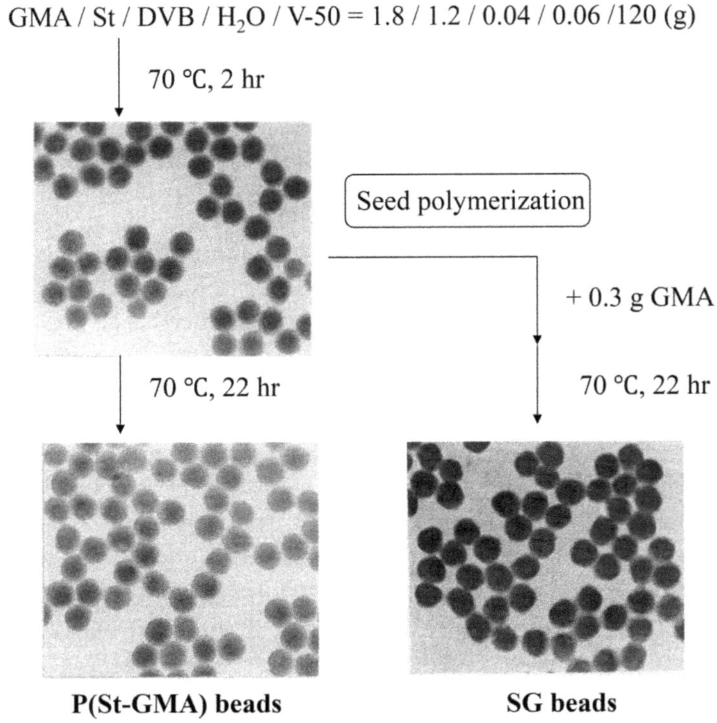

Figure 3.2. TEM views of poly (St-GMA) and SG beads.

that the individual beads had a diameter of ~200 nm and existed as a single-bead dispersion.

3.3 Immobilization of DNA to SG beads

We attempted to purify the transcriptional regulatory factor ATF/E4TF3, which specifically binds the DNA sequence 5′-TGACGTAA-3′ and activates transcription. For this purpose, we generated two oligonucleotides, 5′-AAGTGACGTAACG-TTGGGGGGGGGG-3′ and 5′-AACGTTACGTCATTCCCCCCCCCC-3′, which were annealed to form a duplex. Thereafter, both 5′ ends were phosphorylated, and the duplex was ligated, yielding DNA chains with an average length of 300 base pairs, with sticky 3′ ends (figure 3.3).

Immobilization of DNA on SG beads was performed in two ways (figure 3.4). In one approach, we used the CNBr method through the diol group by opening epoxy groups with hydrochloric acid, similar to the strategy of Tjian. In the other approach, we used direct coupling between epoxy groups derived from poly GMA and the amino groups of DNA. Figure 3.5 shows the correlation between the amount of DNA on feed and the amount of DNA immobilized onto the beads. The amount of immobilized DNA was higher in the epoxide-direct-coupling method than in the conventional CNBr method. This is likely because the CNBr method caused the vice-reaction of some at a stage to be activated step of diol groups, as shown in figure 3.4(A). In addition, the beads handled by the CNBr method were

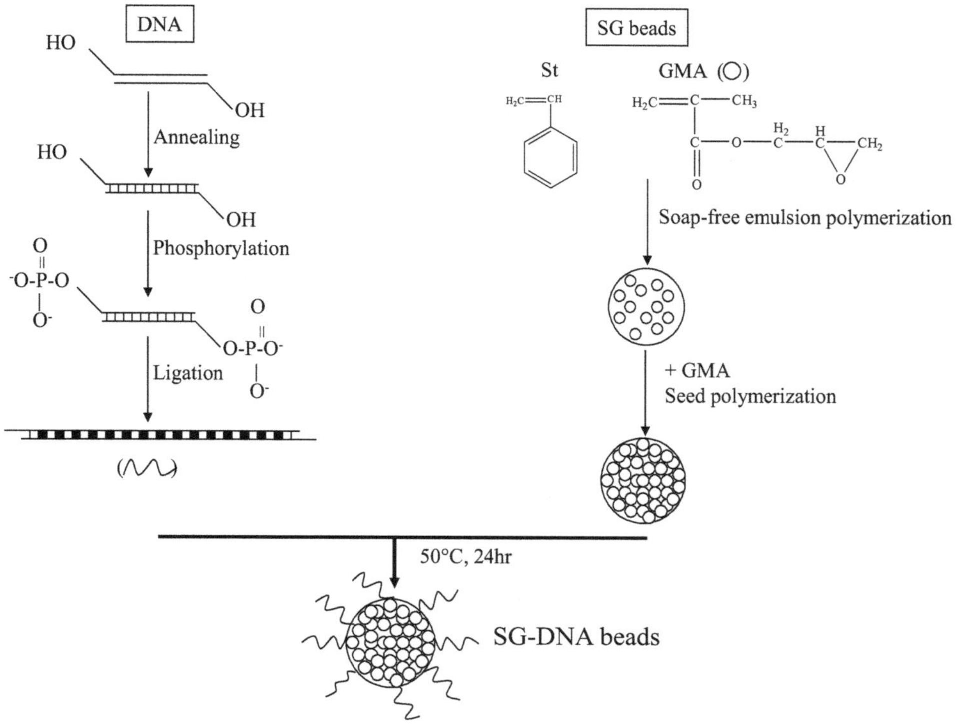

Figure 3.3. Preparation of DNA-immobilized SG beads for affinity purification. Two oligonucleotides, 5′-AAGTGACGTAACGTTGGGGGGGGGG-3′ and 5′-AACGTTACGTCATTCCCCCCCCCC-3′, were synthesized, annealed, phosphorylated, and ligated. Double-stranded DNA fragments ~300 bp in length, with sticky 3′ ends, were obtained for immobilization onto beads.

slightly unstable, and tended to aggregate after activation with CNBr; consequently, the amount of immobilized DNA was reduced. Ultimately, the epoxide method was more efficient than the CNBr method.

3.4 Purification of a transcriptional regulatory factor bound to a specific DNA sequence

In this experiment, 2.5 mg (solid content) of DNA-immobilized beads (SG-DNA) in a 1.5 mL microtube was washed three times in 50 mM Tris–HCl [pH 7.9], 20% (v/v) glycerol, 0.1 M KCl, 1 mM EDTA, 1 mM DTT, and 0.1% (w/v) NP-40. After 200 μL of the fraction containing ATF/E4TF3 was mixed with 10 μg of poly(dI-dC) for 15 min, ATF/E4TF3 was bound to DNA immobilized on SG beads by addition of this solution to SG-DNA. After 30 min, the mixture was centrifuged; the supernatant was removed; the beads were washed with 50 mM Tris–HCl [pH 7.9], 20% (v/v) glycerol, 0.1 M KCl, 1 mM EDTA, 1 mM DTT, 0.1% (w/v) NP-40, and then with 50 mM Tris–HCl [pH 7.9], 20% (v/v) glycerol, 1.0 M KCl, 1 mM EDTA, 1 mM DTT, 0.1% of (w/v) NP-40; and ATF/E4TF3 was eluted with 50 mM Tris–HCl [pH 7.9], 20% (v/v) glycerol, 1.0 M KCl, 1 mM EDTA, 1 mM DTT, and 0.1% (w/v) NP-40. In addition, to avoid inactivation of the protein, all the operations

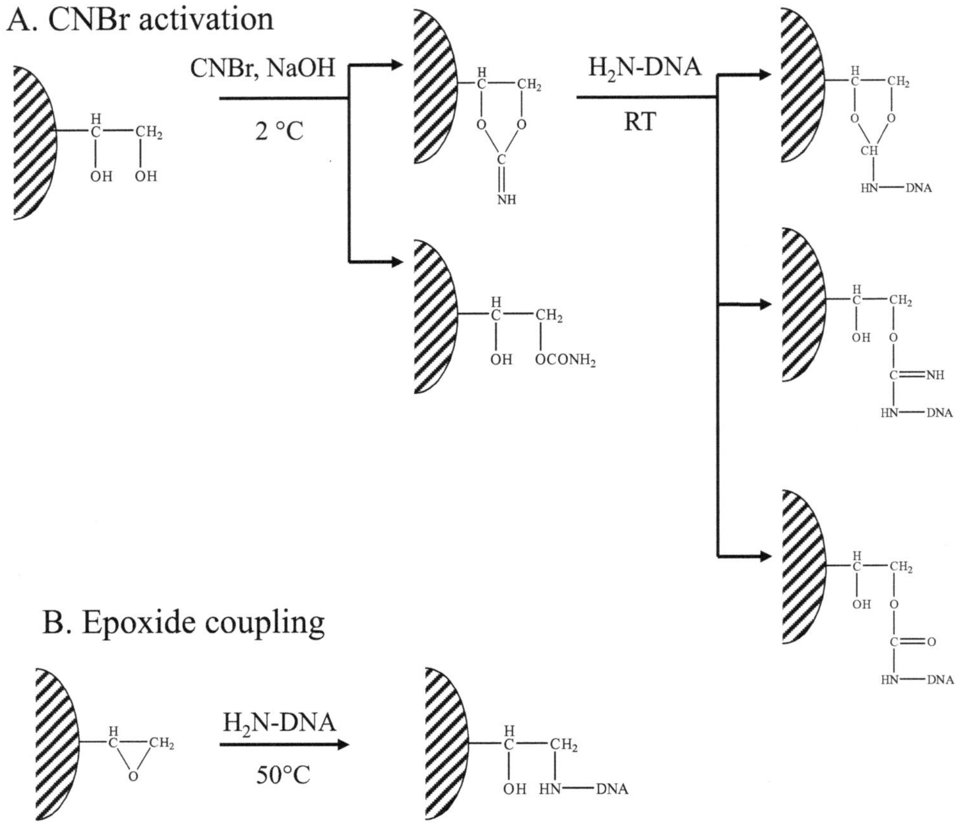

Figure 3.4. Immobilization of DNA onto beads. (A) CNBr activation method. (B) Epoxide-coupling method.

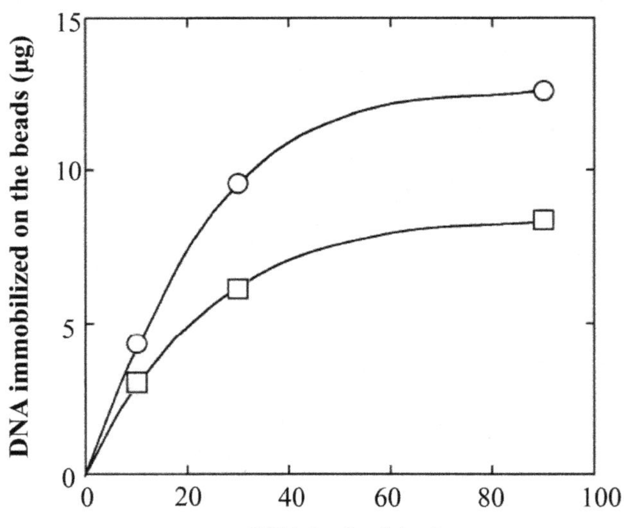

Figure 3.5. Immobilization of DNA onto SG beads by the CNBr (□) and epoxide-coupling (○) methods.

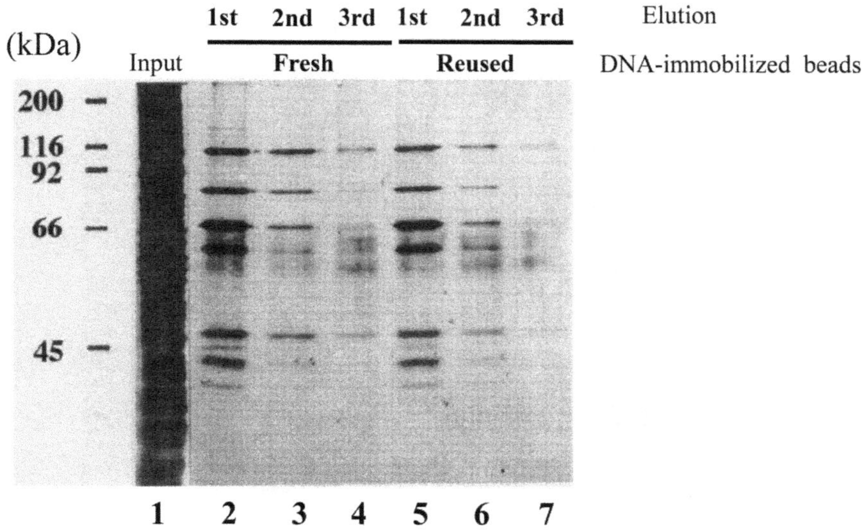

Figure 3.6. Identification of proteins in fractions after purification of ATF/E4TF3. Aliquots of affinity-purified fractions were subjected to 10% SDS-PAGE. The gel was silver-stained to visualize the proteins.

of separation and purification were performed at 4 °C, and the eluate was stored at −80 °C.

Fractions were analyzed by SDS-PAGE and silver staining (figure 3.6). Lanes 2–4 show proteins eluted with 1.0 M KCl Tris buffer. Polypeptides with a molecular weight of 80, 66, 60, 47, 43, and 39 kDa were detected in the eluate, and were specifically bound to DNA. Furthermore, if the SG-DNA beads were washed three times with 50 mM Tris–HCl [pH 7.9], 20% (v/v) glycerol, 2.5 M KCl, 1 mM EDTA, 1 mM DTT, and 0.1% (w/v) NP-40 and dispersed with 10 mM Tris–HCl [pH 8.0], 0.3 M NaCl, 1 mM EDTA, and 0.02% (w/v) NaN3 at 4 °C for 1 week, it was possible to regenerate and reuse these SG-DNA beads, as shown in lanes 5–7.

3.5 Purification from crude cell extract with SG-DNA beads

In section 3.4, ATF/E4TF3 was purified using SG-DNA beads from a nuclear extract preliminarily crudely purified with heparin–Sepharose and DEAE-Sepharose. Subsequently, direct affinity purification of ATF/E4TF3 was performed by SG-DNA beads from the nuclear extract. Figure 3.7 confirms that a 43 kDa protein corresponding to the ATF/E4TF3 family was obtained. Ultimately, 2 μg of the 43 kDa polypeptide was recovered at a concentration of 6.0 mg mL^{-1} from 1 mL of nuclear extract.

The yield of this purification was approximately 20 times greater than that of conventional DNA-immobilized agarose. In addition, in contrast to conventional DNA-immobilized Sepharose, the operation could be easily performed by the batch-wise method in only 2 h [5].

Finally, we examined this purified fraction to determine whether it was functional in an *in vitro* transcription system (figure 3.8). The fraction treated with SG-DNA

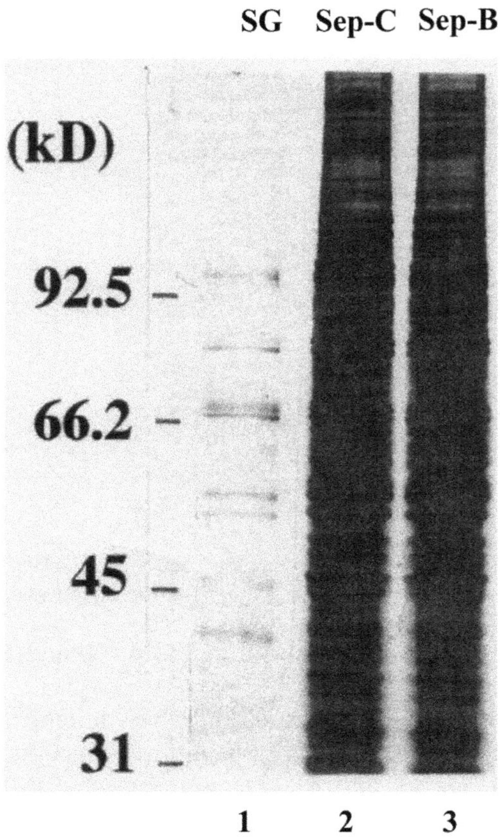

Figure 3.7. Comparison of the purification efficiency of DNA-immobilized SG beads (SG-DNA beads) and Sepharose resins. Aliquots of fractions affinity-purified with SG-DNA beads or DNA-immobilized Sepharose resins were determined by SDS-PAGE. SG beads: eluates from batch-wise purification with SG-DNA beads; Sep-C and Sep-B: column chromatography with the indicated DNA-immobilized Sepharose resins, respectively.

beads (flow-through) exhibited reduced transcriptional activity, dependent on ATF/E4TF3 (lanes 1 and 2). However, transcriptional activity was restored to the original state when the fraction purified by SG-DNA beads was added (lane 4). These results confirm that ATF/E4TF3 is a positive factor that promotes transcription, as previously reported, and that these SG-DNA beads can be applied to *in vitro* transcription systems [4].

Analysis of the transcription factor groups can be performed actively by the research achievement of these SG-DNA beads—though the purification was difficult until now—and it significantly contributes to the elucidation of the molecules' level of transcription control. In the future, we expect that our beads will be applied not only to transcription factors, but also to general DNA-binding proteins, and will greatly contribute to the development of this research field.

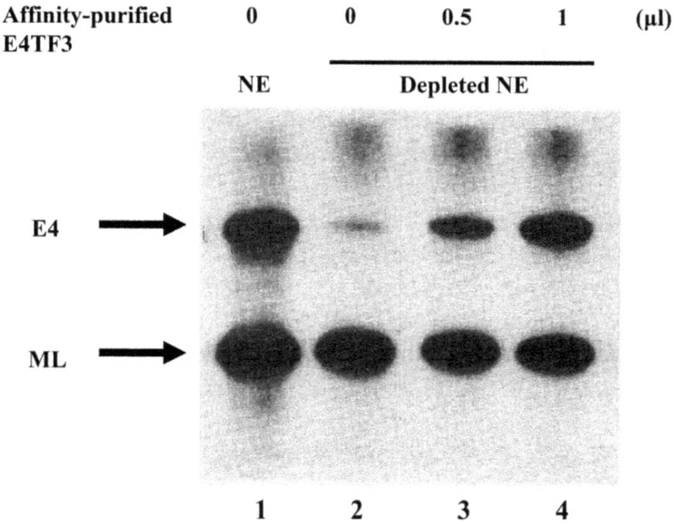

Figure 3.8. Transcriptional activity of affinity-purified ATF/E4TF3. The reactions were incubated at 30 °C for 60 min, and RNA transcripts were extracted and analyzed on 6% urea–polyacrylamide gels. Templates contained the adenovirus E4 and major late (ML) promoters and G-free cassettes of different lengths. The products, E4 and ML, were distinguishable based their size (400 and 270 nucleotides, respectively).

3.6 Development of FG beads

SG beads enabled a shift from the column to the batch method, allowing quick, high purity, high-yield recovery of the target protein from cell crude extract. However, in affinity purification using SG beads, all procedures must be performed by manual labor, requiring a great deal of effort and time. Therefore, we sought to automate the affinity purification step. For this purpose, we developed highly functionalized magnetic beads (FG beads), with characteristics superior to those of SG beads and that were responsive to magnetism.

In collaboration with Professor Masanori Abe, Graduate School of Science and Technology, Tokyo Institute of Technology, we found that acidic amino acids preferentially bind to ferrite (magnetic iron oxide). In particular, carboxyl and thiol groups bind strongly to the surface of ferrite particles [9]. Therefore, in addition to amino acids, compounds with functional groups, such as carboxyls and thiols, could be used to modify and process the surfaces of ferrite particles. We refer to these compounds as 'adapter molecules'. We searched for the most suitable adapter molecules for the formation of polymeric beads encapsulating ferrite particles [10]. Ultimately, we identified 10-undecenoic acid as an optimal molecule for this purpose. This compound has both carboxyl groups for binding to the surface of ferrite particles and a double bond that can easily react with other monomers, facilitating polymerization. We prepared magnetic iron particles 35–40 nm and coupled 10-undecenoic acid to the particles as an adapter molecule. Through the adapter molecules, the magnetic iron particles were coated with a copolymer with St and GMA, and then completely coated by seed polymerization to generate poly

GMA. As shown in figure 3.9, in the resulting particles (called FG beads), the ratio of Poly-St was increased, bringing the shape closer to a true sphere [11].

The FG beads were polymeric particles approximately 200 nm in diameter, and TEM observation revealed that each bead contained several ferrite nanoparticles. In general, extremely large or small particles were not observed, and the size distribution of the beads was within 20% of the average diameter. FGNEGDE beads, modified on the surface with ethylene glycol diglycidyl ether (EGDE), could be dispersed in various organic solvents, e.g. DMF, THF, ethyl acetate, 1,4-dioxane, toluene, and dichloromethane, and could be subsequently dispersed in water without being destroyed by the organic solvents. As a result, FGNEGDE beads were much more stable in various organic solvents, and it was clear that it would be possible to immobilize the ligand on the surface of FGNEGDE beads in suitable solvents.

These FG beads were much more tolerant of organic solvent than SG beads, as shown in figure 3.10. As a result, we could immobilize compounds that are insoluble in water on the surface of beads, and it came to be thereby possible for the immobilization of the compounds if they had even functional groups, though it was impossible for the immobilization of compounds that were insoluble in water, as ever.

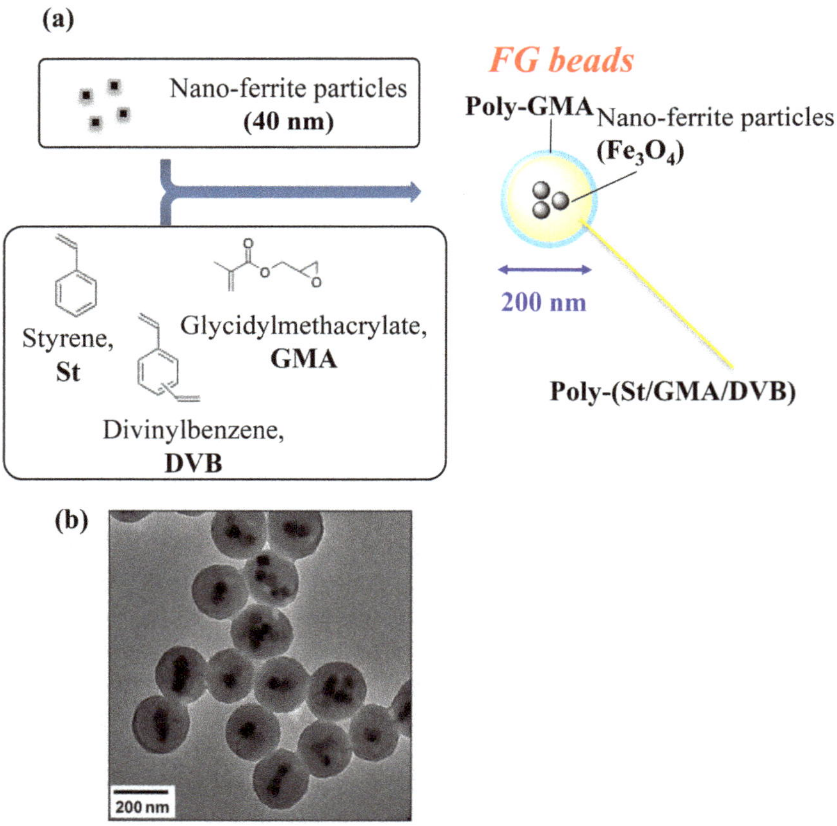

Figure 3.9. FG beads. (a) Preparation of FG beads. (b) TEM view of FG beads.

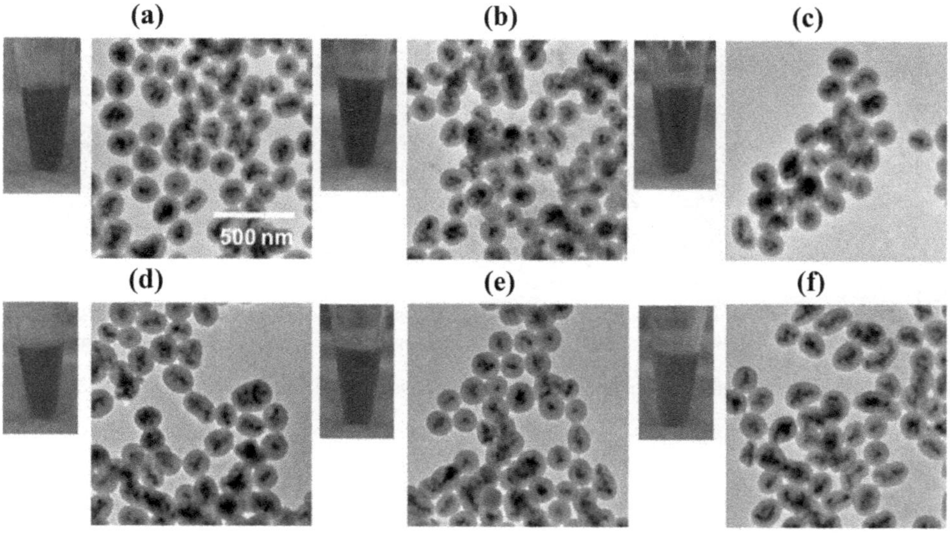

Figure 3.10. Suspension of FGNEGDE beads in organic solvents, and TEM views of FGNEGDE beads from each organic solvent resuspended in water. (a) *N*,*N*-dimethylformamide (DMF), (b) tetrahydrofuran (THF), (c) ethyl acetate, (d) 1,4-dioxane, (e) toluene, (f) dichloromethane.

Automated screening equipment [12] and FG beads are commercially available from Tamagawa Seiki Co. Ltd. 'FG beads' is a registered trademark (figure 3.11). Today, beads with various functional groups, e.g. carboxyl, NHS-activated, tosyl-activated, and azido, are already on the market.

In the sections below, we introduce various examples of isolation of small molecules using SG beads or FG beads. One of the most successful cases of target protein isolation using FG beads is the discovery of cereblon (CRBN), a primary target of thalidomide. Therefore, a comparatively large amount of text is devoted to the discovery of CRBN and its application.

3.7 Identification of the direct targets of internal bioactive compounds

The cells of our body require many compounds to maintain their metabolism and life processes. Small molecules such as vitamins and amino acids bind to specific target proteins. However, the target proteins of these internal molecules remain largely unknown. One class of these molecules, porphyrins, are composed of four pyrrole rings and are mainly involved in energy metabolism. Heme (figure 3.12(A)), a derivative of porphyrin (Fe-Protoporphyrin IX, figure 3.12(B)), plays an important role in various biological processes such as oxygen transport, respiration, and detoxification. Heme functions as a co-factor for several hemoproteins [13]. For more than a decade, we have attempted the isolation of new porphyrin- or heme-binding proteins using SG or FG beads, with the goal of revealing new aspects of bioregulation.

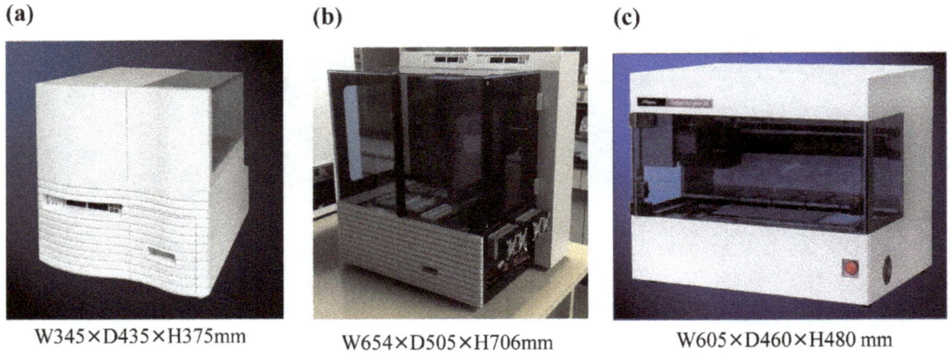

Figure 3.11. Automated screening equipment (bio screening systems). (a) For 8 samples. (b) For 24 samples. (c) For 96 samples.

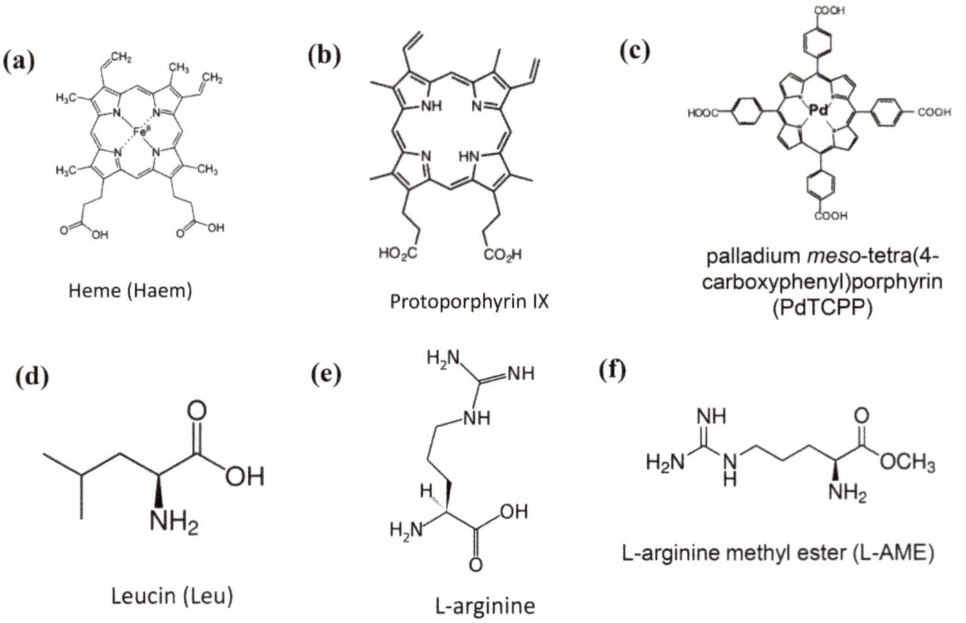

Figure 3.12. Structure of internal bioactive compounds whose target proteins were isolated by FG beads. (a) Heme; (b) Protoporphyrin IX; (c) PdTCPP; (d) Leucin; (e) L-arginine; (f) L-AME.

Palladium meso-tetra(4-carboxyphenyl)porphyrin (PdTCPP) (figure 3.12(C)), a phosphorescent porphyrin derivative, accumulates in mitochondria [14], but the mechanism of accumulation was previously unknown. To investigate this issue, we performed target isolation using PdTCPP-immobilized beads. PdTCPP was first mixed with N-hydroxysuccinimide (NHS) and transformed to succinated PdTCPP. Then, succinated PdTCPP molecules were coupled to the amino groups of SG beads to achieve immobilization. Through this approach, an oxoglutarate carrier (OGC) was identified as a new PdTCPP-binding protein [15]. OGC, a transporter in the inner membrane of mitochondria, transfers 2-oxoglutarate into mitochondria via electroneutral exchange for malate. Our results demonstrated that OGC is a novel carrier of porphyrin that transfers this small molecule into mitochondria. We concluded that porphyrin and derivatives, including PdTCPP, accumulated via the activity of OGC. Previous studies showed that porphyrins tend to be retained in tumors and cancer cells. Thus, these findings contribute to our understanding of the physiological significance of porphyrin accumulation in tumors.

Heme is produced by ferrochelatase (FECH), which catalyzes the insertion of an iron ion into Protoporphyrin IX (PPIX), a precursor of heme, on the inner surface of mitochondria. Subsequently, heme is utilized in the mitochondrial matrix as a co-factor of hemoproteins such as cytochromes. Accumulation of PPIX in the mitochondria is important for the maintenance of bioregulation, but the mechanism was largely unknown. Hence, we performed protein isolation from rat mitochondrial extracts using heme- or PPIX-immobilized beads, and identified adenine nucleotide translocator 2 (ANT2) as a common direct target of both PPIX and heme. Biochemical and genetic analyses showed that ANT contributes to the transport of the heme precursor into the mitochondrial matrix [16].

Progesterone receptor membrane component 1 (PGRMC1) was recently identified as a new heme-binding protein [17]. PGRMC1 belongs to the membrane-associated progesterone receptor (MAPR) family and has a cytochrome b5-like heme-binding domain. Heme dimerizes PGRMC1. Carbon monoxide (CO) binds to heme, thereby inhibiting dimerization of PGRMC1. X-ray analyses showed that a molecule of heme is inserted to the dimeric PGRMC1 molecules. In addition, dimerization enables PGRMC1 to bind to EGFR and cytochrome P450, resulting in the promotion of cancer growth and conferring chemoresistance against anticancer agents. Thus, this study provides a novel insight into the role of PGRMC1 in cancer.

Our cells contain 20 amino acids, which are the building blocks of proteins (polypeptides) and peptide hormones. However, the roles of individual amino acids have also interested many investigators. It has been known for decades that leucine (Leu) (figure 3.12(D)) can stimulate protein synthesis in cells. Because the underlying mechanism was unknown, we sought to identify a novel target of leucine. To prepare Leu-immobilized beads, F-moc-Leu-OH was first conjugated to FG beads, and then the F-moc was removed by piperidine. UBR1 (ubiquitin-protein ligase E3 component n-recognin 1) and UBR2 was identified as a Leu-binding protein present in leucine-responsive rat hepatoma cells [18]. UBR1 and UBR2 are E3 ubiquitin ligases responsible for degradation by the N-end rule. An E3 ubiquitin ligase is an enzyme that catalyzes the transfer of ubiquitin from an E2 ubiquitin-conjugating enzyme to its specific substrate(s) in the ubiquitin-proteasome system. Ubiquitinated substrates undergo proteolysis mediated by the proteasome, a huge protease complex. The N-end rule is the system that controls protein degradation based on the identities of the N-terminal amino acids of protein substrates. UBR1 and UBR2 recognize N-terminal residues and induce selective degradation of target proteins according to this rule. *In vitro* studies show that Leu binds the substrate-recognition domain of UBR2 and thereby inhibits degradation of the N-end rule substrates. Knockdown of UBR1 and UBR2 activates the protein synthesis pathway regulated by mTOR.

Therefore, because UBR1 and UBR2 negatively regulate the mTOR-based protein synthesis pathway, and Leu inhibits the activity of UBR1 and UBR2, Leu may activate protein synthesis.

We have also been investigating the direct targets of other amino acids. One of them, arginine, is a charged amino acid that is important for various aspects of biological regulation including cell division, wound healing, removal of ammonia, and release of hormones. In addition, arginine treatment promotes insulin production in mouse pancreatic NIT cells. Identifying arginine (figure 3.12(D))-interacting factors (AIFs) is essential for elucidating the mechanism underlying this process. We found that L-arginine methyl ester (AME; figure 3.12(E)) also induces insulin production; because AME is technically easier to immobilize on beads than arginine, we prepared AME-immobilized beads. Using these beads, we identified RuvB-like 1, RuvB-like 2, and phosphofructokinase as AIFs. The functional relationship between arginine/AME and these binding proteins is currently under investigation [19].

3.8 Identification of the direct targets of natural products

Capsaicin (figure 3.13(A)), a well-known bioactive compound present in chili peppers, has multiple biological activities. For example, it is responsible for the pungent taste of chilis. This compound functions as an agonist of transient receptor potential cation channel subfamily V member 1 (TRPV1) [20], which is involved in the detection of body temperature. Capsaicin binds to TPRV and induces sensations of excessive heat or abrasive damage. In addition to its TRPV1-based activity, capsaicin has anticancer activity. Affinity purification using capsaicin-immobilized beads identified prohibitin 2 (PHB2) as a novel capsaicin-binding protein [21]. PHB2 is normally located in the inner membrane of mitochondria, but capsaicin induces its translocation to the nucleus, where it functions as a pro-apoptotic factor. This study

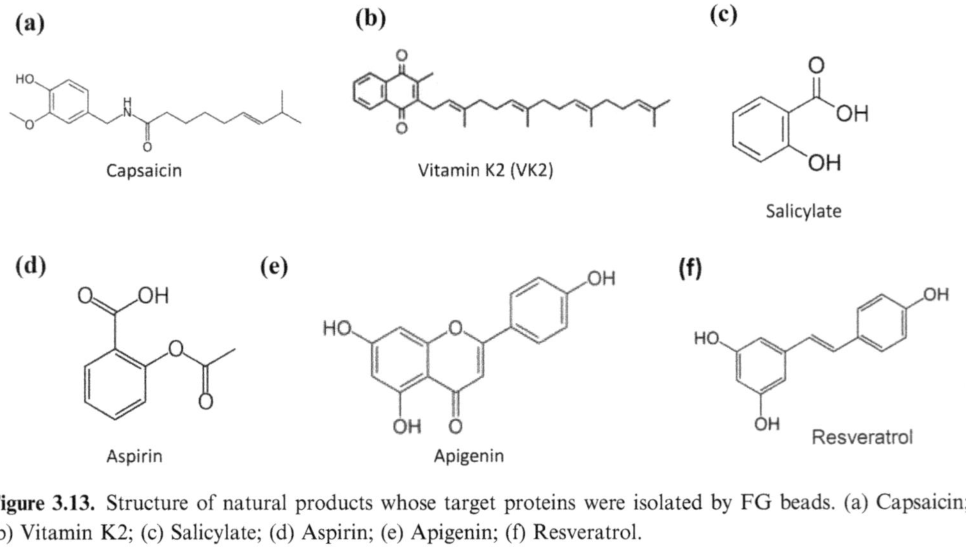

Figure 3.13. Structure of natural products whose target proteins were isolated by FG beads. (a) Capsaicin; (b) Vitamin K2; (c) Salicylate; (d) Aspirin; (e) Apigenin; (f) Resveratrol.

revealed a novel target protein of capsaicin other than TRPV1, and provided insight into the molecular mechanism of capsaicin-induced cell death in cancer cells.

Vitamin K (VK) is a fat-soluble vitamin. Humans are not able to produce VK by themselves, but must acquire it from other organisms consumed as food. VK was originally discovered as a blood coagulation factor. Three types of VK exist: VK1, VK2, and VK3. VK2 (figure 3.13(B)) is a co-factor of γ-glutamyl carboxylase (GGCX) [22], which catalyzes the γ-carboxylation of several substrates including blood coagulation factors. VK2 also binds SXR, a nuclear receptor, and promotes its transcriptional activity [23]. In addition, VK exerts anticancer activity. Previous studies show that VK2 inhibits proliferation of several leukemia and hepatocellular carcinoma cells. To elucidate the anticancer mechanism of VK2, an amino derivative VK2 was succinated with NHS and covalently conjugated to the carboxy group on FG beads. As a result, BAK (Bcl-2 antagonist killer 1) was identified as a novel VK2-binding protein [24]. BAK is a pro-apoptotic factor and a member of the Bcl-2 family. Interestingly, biochemical studies revealed that VK2-2,3 epoxide, a metabolite of VK2, covalently binds to BAK, and this VK2ylated BAK promotes apoptosis. These findings elucidate, at least partly, the anticancer activity of VK2, and may contribute to the development or improvement of anticancer drugs targeting BAK and related factors.

Since ancient times, willow bark has been known to have antipyretic effects. Salicylate (figure 3.13(C)) has been as the active ingredient in willow bark. Chemically synthesized salicylate is now available as the parental compound of the nonsteroidal anti-inflammatory drug aspirin (acetylsalicylate; figure 3.13(D)), which is used worldwide. Although salicylate and aspirin have antipyretic effects, the drugs have side effects such as hepatic toxicity, the mechanism of which remains unclear. Affinity purification using salicylate-immobilized beads identified ferrochelatase (FECH). As described above in the section about heme, FECH catalyzes the conversion of PPIX into heme by adding a Fe ion in the final step of heme synthesis [25]. Salicylate inhibits the enzymatic activity of FECH. To determine whether FECH inhibition by salicylate is involved in the drug's adverse effects, we adopted zebrafish as an animal model. Salicylate inhibited heme synthesis similar to the knockdown of FECH. Conversely, overexpression of FECH reversed the effects salicylate treatment. We concluded that FECH inhibition by salicylate contributes to the drug's adverse effect. These findings may contribute to developing new salicylate derivatives without deleterious side effects.

Flavones, a class of flavonoids contained in vegetables and fruits, possess useful bioactivities including antitumor effects. However, little is known about the precise molecular mechanisms of these natural compounds. Iizumi and colleagues have been investigating the molecular targets of flavones. Their group recently identified ANT2 and ribosomal protein S9 (RPS9) as direct targets of the flavone apigenin (figure 3.13(E)), which has antiproliferative effects in several types of cancer cells [26, 27]. ANT2 was isolated from prostate cancer cells. Binding of apigenin to ANT2 results in upregulation of the death receptor 5 and enhancement of apoptosis mediated by Apo2/TRAIL. RPS9 was isolated from colon cancer cells. Binding of apigenin to RPS9 results in downregulation of cyclin-dependent kinase 1 (CDK1). Apigenin induces G2/M arrest in several cancer cell lines, and this arrest is mediated at least partially by the binding of RPS9.

Resveratrol (figure 3.13(F)) is a well-known bioactive compound isolated from *Veratrum album* by a Japanese researcher in 1939. Numerous reports show that resveratrol has multiple biological activities, including anticancer and anti-aging effects. The precise mechanism of the action of resveratrol remains elusive. SIRT1 (sirtuin 1), an NAD-dependent deacetylase responsible for health and longevity, is activated by resveratrol [28]. However, it remains possible that SIRT1-independent pathways are involved in the mechanism of resveratrol action. Target screening using FG beads recently showed that the DEAD (Asp-Glu-Ala-Asp) box helicase 5 (DDX5) is a novel target of resveratrol [29]. DDX5, which is highly expressed in several types of cancer cells, promotes proliferation and metastasis, and also confers drug resistance mediated by activation of several molecular pathways. The screening study showed that resveratrol binds to DDX5, resulting in suppression of mTORC1 signaling (mammalian target of rapamycin complex 1) in several prostate cancer cell lines.

Target isolation of natural bioactive compounds may provide new 'druggable' targets. The term 'druggability' is defined as the ability of a biological target, such as a protein, to bind and be modulated by small chemicals [30]. The targets of a huge number of natural bioactive compounds are unknown, and identification of these targets might reveal unexpected druggable proteins, leading to the development of new drugs.

3.9 Identification of the direct targets of pharmaceutical drugs

E3330 (figure 3.14(A)), developed as an anti-inflammatory drug by Eisai Co, Ltd, suppresses the activation of the NF-κB pathway. NF-κB is a well-known transcription factor that induces production of multiple proteins associated with

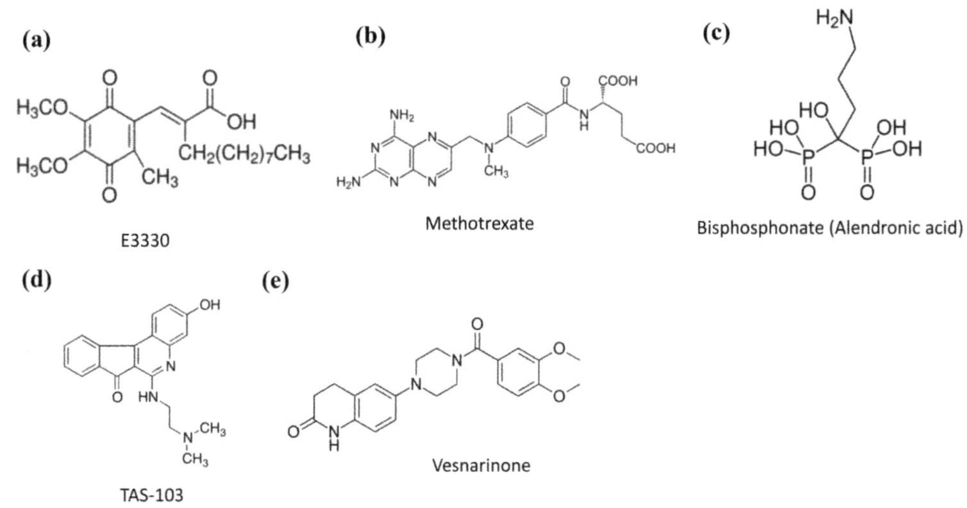

Figure 3.14. Structure of pharmaceutical drugs whose target proteins were isolated by FG beads. E3330; (b) Methotrexate; (c) Bisphosphonate; (d) TAS-103; (e) Vesnarinone.

inflammatory responses. Usually, NF-κB forms a protein complex with I-κB, which inhibits its activation. Certain stimuli such as treatment with TNF-alpha induce degradation of IkB, resulting in activation of NF-κB. However, E3330 does not affect IκB degradation, which suggests that it affects an IkB-independent process. Hence, elucidation of its mechanism of action would provide deeper insight into the NF-kB pathway. Therefore, we immobilized E3330 on SG beads and purified E3330-binding proteins. This analysis identified Ref-1, a transcription factor involved in redox changes, as a drug-binding protein [30]. E3330 inhibits Ref-1 activity, which promotes DNA-binding of NF-κB p50, resulting in the repression of NF-κB-induced transcriptions. These findings suggested that Ref-1 represents a potential new therapeutic target for anti-inflammatory drugs.

Methotrexate (MTX; figure 3.14(B)) is a well-known anticancer drug that blocks nucleotide synthesis by inhibiting dihydrofolate reductase (DHFR). We prepared two types of MTX-immobilized beads: in one case, the beads were conjugated to the γ-carboxylate group of MTX (MTX beads-1), and in the other case, the beads were conjugated to another part of the molecule (MTX beads-2). MTX beads-1 bound DHFR (23 kDa), as previously reported. On the other hand, MTX beads-2 bound to a 30 kDa protein subsequently identified as deoxycytidine kinase (dCK), a novel MTX-binding enzyme involved in the salvage pathway of nucleotide synthesis [31]. MTX is often used in combination with 1-β-D-arabinofuranosylcytosine (ara-C), a chemotherapeutic agent. dCK phosphorylates ara-C, resulting in its incorporation into chromosomal DNA. We found that MTX directly bound to dCK and promoted the incorporation and cytotoxicity of ara-C in Burkitt's lymphoma cells. Thus, this study revealed a new mechanism of MTX. Furthermore, this experiment provides a good example of the fact that target isolation using SG or FG beads sometimes depends on the immobilization position.

Bisphosphonates are well-known osteoporosis drugs. Although these compounds are prenylation inhibitors, and are thought to exert their effects on bone resorption by disrupting signaling downstream of prenylated small GTPases, the molecular mechanism remained to be fully elucidated. Unexpectedly, we found that bisphosphonate (alendtronic acid; figure 3.14(C)) inhibited the Simian Virus 40 (SV40) and adenovirus infection via a pathway independent of prenylation. To investigate this finding, we performed affinity purification using bisphosphonate-immobilized beads, and isolated Dynamin-2 as a direct target [32]. We then showed that bisphosphonate blocks endocytosis by inhibiting the GTPase activity of Dynamin-2. This study uncovered an unknown aspect of the mechanism of bisphosphonate activity, raising the possibility of developing new anti-viral drugs targeting Dynamin-2.

TAS-103 (figure 3.14(D)), a quinoline derivative that inhibits topoisomerase II, possesses antitumor activity. Nevertheless, cellular sensitivity to TAS-103 is not correlated with the expression level of topoisomerase II. Therefore, TAS-103 is likely to have another target. TAS-103-immobilized beads purified a signal recognition particle (SRP) [33], a ribonucleoprotein complex composed of six proteins (SRP9, SRP14, SRP19, SRP54, SRP68, and SRP72) and SRP RNA. SRP54 recognizes the signal peptides of membrane and secretory proteins as they emerge from the ribosome. After recognition, SRP delivers the nascent protein to the endoplasmic reticulum (ER).

TAS-103 disrupted the SRP, resulting in the accumulation of secreted proteins such as IL-6. This study provided new insights into the mechanism of action of TAS-103.

Vesnarinone (figure 3.14(E)), a synthetic quinolinone derivative developed for the treatment of cardiac failure and certain types of cancers, inhibits production of TNFα. Previously, however, the mechanism of this action was unknown. Using FG beads, a valosin-containing protein (VCP) was identified as a vesnarinone-binding protein [34]. VCP is a member of the triple ATPases family (AAA), which plays an important role in many cellular processes such as endoplasmic reticulum-associated degradation of proteins (ERAD), cell cycle, and DNA repair. In addition, VCP is involved in the ubiquitin proteasome: specifically, VCP binds to the polyubiquitin chains of certain substrate proteins (e.g. IκB) and recruits them to 26 S proteasome, resulting in proteolysis. Vesnarinone was shown to disrupt the interaction between VCP and IκB, resulting in accumulation of IκB. As noted above, accumulation of the IκB protein inhibits the activation of NFκB. Thus, vesnarinone decreased the expression of TNFα, a target of NF-κB activation. These results indicate that vesnarinone inhibits the NF-κB pathway by disrupting the interaction between VCP and IκB.

3.10 Identification of a direct target of thalidomide using FG beads

In the 1950s, thalidomide (figure 3.15(A)) was developed as a sedative drug by the West German pharmaceutical company Grunenthal. Thalidomide consists of a

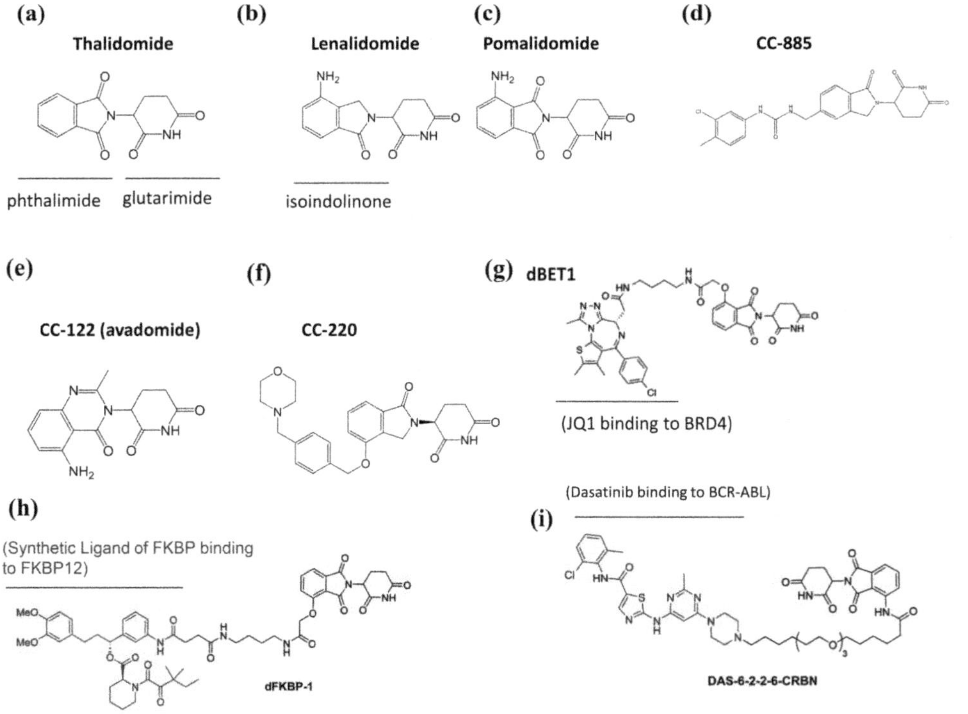

Figure 3.15. Structure of thalidomide and its derivatives. (a) Thalidomide; (b) Lenalidomdie; (c) Pomalidomide; (d) CC-885; (e) CC-122; (f) CC-220; (g) dBET1; (h) dFKBP-1; (i) DAS-6-2-2-6-CRBN.

glutarimide ring with a phthalimido substituent. When this drug is administered to rodents, no apparent phenotype is observed. Consequently, the drug was recognized as safe and sold in over 40 countries. In the 1960s, however, researchers realized that thalidomide is teratogenic: when women took this drug during early pregnancy, their babies had multiple birth defects, including malformations of the limb and ear [35]. Although thalidomide was withdrawn from the market, investigators have continued to study its mechanism of action. Recently, thalidomide was re-evaluated as a clinically effective drug for the treatment of erythema nodosum leprosum (a type of leprosy) and multiple myeloma (a type of B-cell malignancy). In addition, two potent thalidomide derivatives, lenalidomide (figure 3.15(B)) and pomalidomide (figure 3.15(C)), were developed. These compounds, considered immunomodulatory drugs (IMiDs), are used for the treatment of multiple myeloma and other diseases [36]. Nevertheless, the direct target of thalidomide and IMiDs had remained unclear until recently.

We prepared thalidomide-immobilized FG beads to isolate the drug's direct target(s). The carboxy-type thalidomide derivative FR259625 was covalently attached to an amino group of the beads, and the thalidomide beads were mixed with various cell extracts. After extensive washing, the bound proteins were eluted with free thalidomide molecules. Surprisingly, only two protein signals were detected in all cases: the larger protein was DDB1 (damage-specific DNA-binding protein 1), and the smaller one was cereblon (CRBN). Thalidomide binding assays using recombinant proteins showed that only CRBN directly binds thalidomide, whereas DDB1 binds CRBN and thus indirectly interacts with thalidomide.

CRBN was initially reported as a gene related to mental retardation, but its function was largely unknown [37]. By contrast, DDB1 is well-studied: the protein is a subunit of the Cullin-Ring E3 ubiquitin ligase complex (CRL4) [38]. CRL4 is composed of Cullin 4 (Cul4), Roc1 (Regulator of Cullins 1), DDB1, and a substrate receptor that directly recognizes the substrate. Like DDB2, a known substrate receptor of CRL4, CRBN is auto-ubiquitinated. Furthermore, CRBN competes with DDB2 for binding to DDB1, which suggests that it is a novel substrate receptor of CRL4. Auto-ubiquitination of CRBN was inhibited by thalidomide, implying that it is the biochemical target of this drug.

Next, we investigated whether CRBN is involved in the effects of thalidomide. Rodents were not suitable for this purpose because they are resistant to this drug. Therefore, we adopted zebrafish as our main animal model. Zebrafish CRBN is around 70% homologous to human CRBN, and bound both thalidomide and human DDB1. Moreover, thalidomide induced defects in fins, which are counterparts of mammalian limbs. To investigate whether CRBN is a genuine target of thalidomide teratogenicity, we constructed a thalidomide binding deficient mutant $CRBN^{YW/AA}$ (Y384A/W386A). Zebrafish expressing $CRBN^{YW/AA}$ were resistant to thalidomide treatment. We confirmed this result in a chick model. Therefore, we conclude that CRBN is the primary direct target of thalidomide teratogenicity [39].

In parallel with the analysis described above, we sought to determine which part of thalidomide binds CRBN. To this end, we performed a competitive elution assay in which thalidomide beads were mixed with cell lysates, and bound proteins were eluted with glutarimide or phthalimide after washing. CRBN was eluted in the presence of glutarimide, suggesting that the glutarimide moiety is sufficient for binding [40].

Thalidomide has a chiral carbon, and consequently has two optical isomers: *S*- and *R*-thalidomide. A longstanding question is which enantiomer is responsible for the teratogenic effect. We performed the competitive elution assay using both isomers and found that *S*-thalidomide bound more strongly to CRBN than *R*-thalidomide [41].

Subsequently, the x-ray structure of the thalidomide–CRBN–DDB1 complex has been determined by two groups, confirming the results of our biochemical study [41, 42].

3.11 Functions of CRBN, the common direct primary target of IMiDs and other glutarimide-containing compounds

As described above, CRBN binds to the glutarimide moiety of thalidomide. Therefore, it is not surprising that numerous compounds with a glutarimide moiety possess CRBN-binding activity. Among the IMiDs, lenalidomide and pomalidomide also contain a glutarimide. Therefore, we investigated whether CRBN binds to these compounds. The results showed that these two compounds bind to CRBN more strongly than thalidomide. Lenalidomide and pomalidomide have been investigated for treatment of multiple myeloma. Therefore, it would be interesting to determine whether CRBN is involved in the anti-myeloma activity of these compounds. Several researchers performed knockdown experiments using RNAi. In these experiments, lenalidomide and pomalidomide effectively inhibited cell proliferation of myeloma cells, whereas knockdown of CRBN conferred drug resistance to myeloma cells. In addition, several myeloma cell lines that are resistant to lenalidomide or pomalidomide have been established by prolonged exposure to increasing doses of these drugs. Immunoblot analysis showed that CRBN protein is not expressed in several of these cell lines. Consistent with this, some patients receiving long-term lenalidomide therapy acquire resistance to this drug. One research group analyzed samples from patients before and after lenalidomide therapy, and found that *CRBN* mRNA levels were markedly reduced after acquisition of drug resistance. These findings imply that CRBN is required for the anti-myeloma effect of lenalidomide and pomalidomide [40, 43].

As mentioned above, CRBN is a substrate receptor of CRL4, suggesting that several substrates of CRBN are involved in the effects of thalidomide or IMiDs. Three groups independently showed that $CRL4^{CRBN}$ induces protein degradation of Ikaros and Aiolos in the presence of lenalidomide or pomalidomide [44–46]. Both proteins are transcription factors responsible for myeloma survival that repress IL-2 expression in T-cells. Lenalidomide and pomalidomide induce neomorphic alteration of the substrate specificity of $CRL4^{CRBN}$. The degradation of Ikaros and Aiolos by IMiDs clearly explains why IMiDs are effective for the treatment of myeloma. Thalidomide also decreases the levels of Ikaros and Aiolos, although its activity is comparatively weak.

Lenalidomide is effective for the treatment of not only myeloma, but also myelodysplastic syndrome 5q− (5q− syndrome). 5q− syndrome is characterized by partial deletion of one copy of chromosome 5. This haploinsufficiency causes excess proliferation of a certain type of blood cells, concomitant with ineffective differentiation. In comparison with lenalidomide, the effects of thalidomide and pomalidomide on 5q− syndrome are weak. Several groups investigated this mystery. One group identified casein kinase 1a (CK1a) as a specific substrate of $CRL4^{CRBN}$ in the presence of lenalidomide. The *CK1a* gene is located on 5q; consequently, the expression level of Ck1a in 5q− cells is half of that in wild-type cells. Lenalidomide effectively decreased CK1a, whereas the activities of thalidomide and pomalidomide were weaker, suggesting that the extra carbonyl group on the phthalimide of thalidomide or pomalidomide perturbs the interaction between CRBN and CK1a.

The CK1a study addressed an important outstanding question. As mentioned above, rodents are resistant to thalidomide. Notably in this regard, they are insensitive not only to the adverse effects, but also the therapeutic benefits: in mouse cells, neither Ikaros, Aiolos, nor Ck1a are degraded by this family of drugs. Mouse CRBN is more than 90% homologous to human CRBN, but has subtle amino acid differences proximal to the drug-binding pocket. When mouse CRBN was replaced with human in mouse cell lines, surprisingly, the cells became sensitive to lenalidomide. Ultimately, only one amino acid in CRBN (V388) was critical for the recognition of Ikaros, Aiolos, and CK1a. Furthermore, researchers demonstrated that the growth of 'humanized' mouse cells with Ck1a haploinsufficiency was suppressed by lenalidomide. These findings not only revealed the mechanism of lenalidomide action on 5q− syndrome, but suggest that alteration of substrate recognition by $CRL4^{CRBN}$ is dependent on the shape of CRBN ligands (figure 3.16), i.e. certain types of glutarimide-containing compounds [47, 48].

CC-885 (figure 3.15(D)), a novel anticancer agent, was isolated by phenotypic screening of more than 100 cancer cell lines. The compound is similar to lenalidomide, but possesses an extended structure, a urea group and a methyl-chloro-phenyl group. CC-885 has a strong antiproliferative effect against acute myeloid leukemia (AML) cells, but thalidomide and IMiDs have no effect on AML. In a competitive elution assay using thalidomide beads, CC-885 bound to CRBN. Moreover, CC-885 had no effect on AML in which *CRBN* had been ablated by CRISPR/Cas9-based genome editing. Therefore, CC-885, but not other IMiDs, is likely to induce the breakdown of specific substrate(s). GSPT1 (G1 to S phase transition 1) was identified as a new CRBN substrate in the presence of CC-885 [49]. GSPT1 is also known as eRF3a, a subunit of the protein translational termination complex that recognizes STOP codons. CC-885 treatment markedly decreased the abundance of GSPT1 proteins in AML cells, but no other IMiDs (e.g. lenalidomide) had such an effect. By contrast, $GSPT1^{G575N}$ was neither sensitive to, nor bound to, CRBN in the presence of CC-885. AML cells expressing $GSPT1^{G575N}$ were insensitive to CC-885 treatment, implying that the breakdown of GSPT1 drives an antiproliferative effect of CC-885 on AML cells.

X-ray crystallographic studies have revealed the 3D structure of the CRBN/DDB1–CC-885–GSPT1 complex, which shows that CC-885 binds to both CRBN

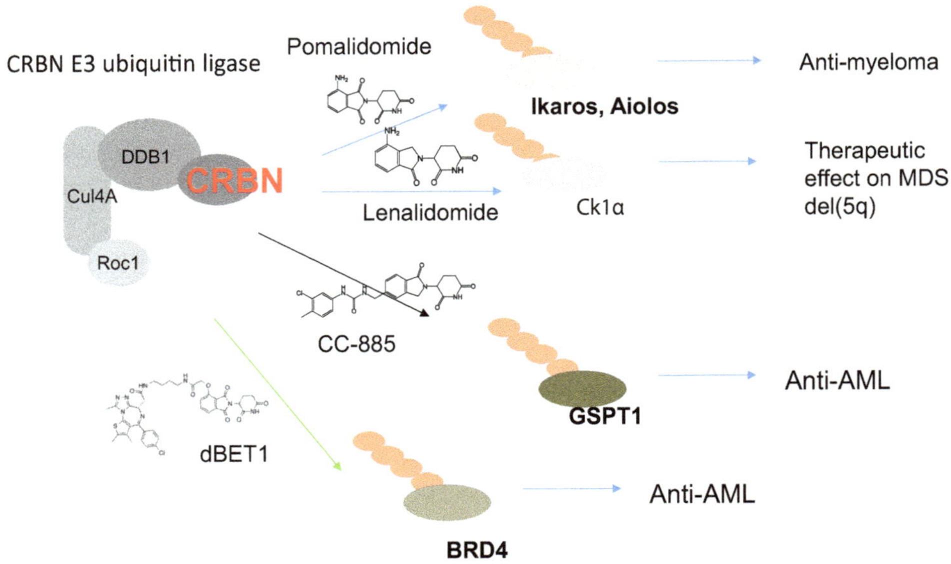

Figure 3.16. Schematic model of CRBN-mediated drug effects. CRBN forms an E3 ubiquitin ligase complex $CRL4^{CRBN}$ with DDB1, Cul4A, and Roc1. $CRL4^{CRBN}$ induced specific protein degradation in a ligand-dependent manner.

and GSPT1, thus providing an interaction hotspot for the two proteins. GSPT1, a substrate of CRBN, contains a specific surface-turn structure containing a glycine (G575) that is critical for binding to CC-885 and CRBN. Glycine, the smallest amino acid, occupies only a tiny area of the interaction hotspot. Consequently, any amino acid substitution at position 575 of GSPT1 abolishes the interaction with CC-885 and CRBN due to steric clash. Ikaros, another substrate of CRBN, also possesses a specific glycine (G151), and the Ikaros G151A mutant is resistant to pomalidomide treatment. There is no apparent sequence homology between GSPT1 and Ikaros, but homology modeling predicts that Ikaros may also contain a surface-turn structure, suggesting that ligand-dependent substrates possess a specific glycine involved in the interaction with ligand and CRBN. These findings provide clues for the development of new CRBN ligands and enable us to rationally identify new substrates for CRBN. Efforts to develop new CRBN ligands for the treatment of AML, based on the CC-885 study, are currently underway [50].

3.12 Development of a new type of CRBN-based drug

Although drug development has been extensively investigated, numerous 'undruggable' protein targets remain. These proteins are responsible for various pathogenic pathways, but they are difficult or impossible to target pharmacologically, preventing chemical modulation or inhibition of these functions. However, induced breakdown of specific proteins by $CRL4^{CRBN}$ makes it possible to indirectly target undruggable proteins such as Ikaros and GSPT1.

The new thalidomide derivatives CC-122 (figure 3.15(E)) and CC-220 (figure 3.15(F)) are currently undergoing clinical trials. An FG bead-based assay confirmed that both compounds bind to CRBN. CC-122 is effective for the treatment of multiple myeloma and diffuse large B-cell lymphoma (DLBCL), and CC-220 is effective against systemic lupus erythematosus (SLE), a type of inflammatory disease [51, 52]. DLBCL is a serious type of malignant non-Hodgkin lymphoma derived from B cells, whereas SLE is a chronic auto-immune disorder that causes severe joint pain and fatigue. The substrates of CRBN responsible for DLBCL or SLE are unknown. Currently, many investigators are trying to identify these substrates. Knowledge of these substrates would provide important clues for developing new CRBN-based drugs targeting DLBCL or SLE.

In parallel with those efforts, new innovative approaches for targeting pathogenic proteins are currently being investigated. In 2015, Bradner and colleagues reported that the new compound dBET1 (figure 3.15(G)), a thalidomide derivative conjugated to a BRD4 inhibitor, induces the degradation of BRD4 proteins by $CRL4^{CRBN}$ [53]. BRD4 is a pathogenic protein required for the cell survival of AML and several types of cancers. The same group also developed dFKBP-1 (figure 3.15(H)), a thalidomide derivative conjugated to an FKBP12-binding compound. Treatment with dFKBP-1 effectively induces the breakdown of FKBP12 proteins. Another group, led by Craig Crews, developed a similar technique called 'Proteolysis Targeting Chimeras' (PROTACs) [54]. This group has achieved the breakdown of BCR-ABL, a pathogenic target of CML, with a new compound consisting of the BCR-ABL inhibitor dasatinib and pomalidomide (DAS-6-2-2-6-CRBN). 'Degraders' such as dBET1 and DAS-6-2-2-6-CRBN (figure 3.15(I)) provide us with opportunities to target numerous pathogenic proteins [55]. Moreover, Bradner and colleagues recently used the BRD4 degraders to study gene expression, and showed that BRD4 is required for a specific step of transcription elongation [56]. Thus, this new technique has already contributed not only to drug development, but also to basic biology.

3.13 Conclusions

Isolation and identification of direct target proteins of small chemicals using SG and FG beads provide preliminary insights into the mechanism of drug action and facilitates the development of new drugs. The targets of a large number of chemicals, including drugs and natural products, remain unknown. Affinity purification using FG beads can contribute to the advancement of a wide range of life science fields, including drug discovery and basic molecular biology. Identification of CRBN, a molecular target of thalidomide, has elucidated the molecular mechanism of this drug's action and facilitated the development of new CRBN-based drugs. Thus, isolation of direct target molecules by affinity purification with FG beads will lead to innovation in drug development.

References

[1] Starkenstein E 1910 *Biochem* **24** 210–8

[2] Kadonaga J T and Tjian R 1986 *Proc. Natl. Acad. Sci. USA* **83** 5889–93
[3] Mitchell P J and Tjian R 1989 *Science* **245** 371–8
[4] Inomata Y, Kawaguchi H, Hiramoto M, Wada T and Handa H 1992 *Anal. Biochem.* **206** 109–14
[5] Inomata Y, Wada T, Handa H, Fujimoto K and Kawaguchi H 1994 *J. Biomater. Sci. Polymer Edn.* **5** 293–302
[6] Kawaguchi H, Asai A, Ohtsuka Y, Watanabe T, Wada T and Handa H 1989 *Nucleic Acids Res.* **17** 6229–40
[7] Inomata Y, Kawaguchi H, Wada T and Handa H 1991 *Koubunshi Ronbunshuu* **48** 335–40
[8] Wada T, Handa H, Inomata Y and Kawaguchi H 1991 *Cell Sci.* **7** 373–81
[9] Nishimura K, Hasegawa M, Ogura Y, Nishi T, Kataoka K, Handa H and Abe M 2002 *J. Appl. Phys.* **91** 8555–6
[10] Nishio K *et al* 2004 *Trans. Mater. Res. Soc. Jpn.* **29** 1659–62
[11] Nishio K *et al* 2008 *Colloids and Surfaces B: Biointerfaces* **64** 162–9
[12] Hanyu N, Nishio K, Hatakeyama M, Yasuno H, Tanaka T, Tada M, Nakagawa T, Sandhu Abe M and Handa H 2009 *J. Magn. Magn. Mater.* **321** 1625–7
[13] Ajioka R S, Phillips J D and Kushner J P 2006 Biosynthesis of heme in mammals *Biochim Biophys Acta* **1763** 723–36
[14] Lo L W, Koch C J and Wilson D F 1996 Calibration of oxygen-dependent quenching of the phosphorescence of Pd-meso-tetra (4-carboxyphenyl) porphine: a phosphor with general application for measuring oxygen concentration in biological systems *Anal. Biochem.* **236** 153–60
[15] Kabe Y, Ohmori M, Shinouchi K, Tsuboi Y, Hirao S, Azuma M, Watanabe H, Okura I and Handa H 2006 Porphyrin accumulation in mitochondria is mediated by 2-oxoglutarate carrier *J. Biol. Chem.* **281** 31729–35
[16] Azuma M, Kabe Y, Kuramori C, Kondo M, Yamaguchi Y and Handa H 2008 Adenine nucleotide translocator transports haem precursors into mitochondria *PLoS ONE* **3** e3070
[17] Kabe Y, Nakane T, Koike I, Yamamoto T, Sugiura Y, Harada E, Sugase K, Shimamura T, Ohmura M and Muraoka K *et al* 2016 Haem-dependent dimerization of PGRMC1/Sigma-2 receptor facilitates cancer proliferation and chemoresistance *Nat. Commun.* **7** 11030
[18] Kume K, Iizumi Y, Shimada M, Ito Y, Kishi T, Yamaguchi Y and Handa H 2010 Role of N-end rule ubiquitin ligases UBR1 and UBR2 in regulating the leucine-mTOR signaling pathway *Genes Cells* **15** 339–49
[19] Hiramoto M, Maekawa N, Kuge T, Ayabe F, Watanabe A, Masaike Y, Hatakeyama M, Handa H and Imai T 2010 High-performance affinity chromatography method for identification of L-arginine interacting factors using magnetic nanobeads 2010 *Biomed. Chromatogr.* **24** 606–12
[20] Caterina M J, Schumacher M A, Tominaga M, Rosen T A, Levine J D and Julius D 1997 The capsaicin receptor: a heat-activated ion channel in the pain pathway *Nature* **389** 816–24
[21] Kuramori C, Azuma M, Kume K, Kaneko Y, Inoue A, Yamaguchi Y, Kabe Y, Hosoya T, Kizaki M and Suematsu M *et al* 2009 Capsaicin binds to prohibitin 2 and displaces it from the mitochondria to the nucleus *Biochem. Biophys. Res. Commun.* **379** 519–25
[22] Stafford D W 2005 The vitamin K cycle *J. Thromb. Haemost.* **3** 1873–8
[23] Tabb M M, Sun A, Zhou C, Grun F, Errandi J, Romero K, Pham H, Inoue S, Mallick S and Lin M *et al* 2003 Vitamin K2 regulation of bone homeostasis is mediated by the steroid and xenobiotic receptor SXR *J. Biol. Chem.* **278** 43919–27

[24] Karasawa S, Azuma M, Kasama T, Sakamoto S, Kabe Y, Imai T, Yamaguchi Y, Miyazawa K and Handa H 2013 Vitamin K2 covalently binds to Bak and induces Bak-mediated apoptosis *Mol. Pharmacol.* **83** 613–20

[25] Gupta V, Liu S, Ando H, Ishii R, Tateno S, Kaneko Y, Yugami M, Sakamoto S, Yamaguchi Y and Nureki O *et al* 2013 Salicylic acid induces mitochondrial injury by inhibiting ferrochelatase heme biosynthesis activity *Mol. Pharmacol.* **84** 824–33

[26] Iizumi Y, Oishi M, Taniguchi T, Goi W, Sowa Y and Sakai T 2013 The flavonoid apigenin downregulates CDK1 by directly targeting ribosomal protein S9 *PLoS One* **8** e73219

[27] Oishi M, Iizumi Y, Taniguchi T, Goi W, Miki T and Sakai T 2013 Apigenin sensitizes prostate cancer cells to Apo2L/TRAIL by targeting adenine nucleotide translocase-2 *PLoS One* **8** e55922

[28] Borra M T, Smith B C and Denu J M 2005 Mechanism of human SIRT1 activation by resveratrol *J. Biol. Chem.* **280** 17187–95

[29] Taniguchi T, Iizumi Y, Watanabe M, Masuda M, Morita M, Aono Y, Toriyama S, Oishi M, Goi W and Sakai T 2016 Resveratrol directly targets DDX5 resulting in suppression of the mTORC1 pathway in prostate cancer *Cell Death Dis.* **7** e2211

[30] Cheng A C, Coleman R G, Smyth K T, Cao Q, Soulard P, Caffrey D R, Salzberg A C and Huang E S 2007 Structure-based maximal affinity model predicts small-molecule druggability *Nat. Biotechnol.* **25** 71–5

[31] Uga H, Kuramori C, Ohta A, Tsuboi Y, Tanaka H, Hatakeyama M, Yamaguchi Y, Takahashi T, Kizaki M and Handa H 2006 A new mechanism of methotrexate action revealed by target screening with affinity beads *Mol. Pharmacol.* **70** 1832–9

[32] Masaike Y, Takagi T, Hirota M, Yamada J, Ishihara S, Yung T M, Inoue T, Sawa C, Sagara H and Sakamoto S *et al* 2010 Identification of dynamin-2-mediated endocytosis as a new target of osteoporosis drugs, bisphosphonates *Mol. Pharmacol.* **77** 262–9

[33] Yoshida M, Kabe Y, Wada T, Asai A and Handa H 2008 A new mechanism of 6-((2-(dimethylamino)ethyl)amino)-3-hydroxy-7H-indeno(2,1-c)quinolin-7-one dihydrochloride (TAS-103) action discovered by target screening with drug-immobilized affinity beads *Mol. Pharmacol.* **73** 987–94

[34] Hotta K, Nashimoto A, Yasumura E, Suzuki M, Azuma M, Iizumi Y, Shima D, Nabeshima R, Hiramoto M and Okada A *et al* 2013 Vesnarinone suppresses TNFalpha mRNA expression by inhibiting valosin-containing protein *Mol. Pharmacol.* **83** 930–8

[35] Miller M T and Stromland K 1999 Teratogen update: thalidomide: a review, with a focus on ocular findings and new potential uses *Teratology* **60** 306–21

[36] Bartlett J B, Dredge K and Dalgleish A G 2004 The evolution of thalidomide and its IMiD derivatives as anticancer agents *Nat. Rev. Cancer* **4** 314–22

[37] Higgins J J, Pucilowska J, Lombardi R Q and Rooney J P 2004 A mutation in a novel ATP-dependent ion protease gene in a kindred with mild mental retardation *Neurology* **63** 1927–31

[38] Lee J and Zhou P 2007 DCAFs, the missing link of the CUL4-DDB1 ubiquitin ligase *Mol. Cell* **26** 775–80

[39] Ito T, Ando H, Suzuki T, Ogura T, Hotta K, Imamura Y, Yamaguchi Y and Handa H 2010 Identification of a primary target of thalidomide teratogenicity *Science* **327** 1345–50

[40] Lopez-Girona A, Mendy D, Ito T, Miller K, Gandhi A K, Kang J, Karasawa S, Carmel G, Jackson P and Abbasian M *et al* 2012 Cereblon is a direct protein target for immunomodulatory and antiproliferative activities of lenalidomide and pomalidomide *Leukemia* **26** 2445

[41] Chamberlain P P, Lopez-Girona A, Miller K, Carmel G, Pagarigan B, Chie-Leon B, Rychak E, Corral L G, Ren Y J and Wang M *et al* 2014 Structure of the human Cereblon-DDB1-lenalidomide complex reveals basis for responsiveness to thalidomide analogs *Nat. Struct. Mol. Biol.* **21** 803–9

[42] Fischer E S, Bohm K, Lydeard J R, Yang H, Stadler M B, Cavadini S, Nagel J, Serluca F, Acker V and Lingaraju G M *et al* 2014 Structure of the DDB1-CRBN E3 ubiquitin ligase in complex with thalidomide *Nature* **512** 49–53

[43] Zhu Y X, Braggio E, Shi C X, Bruins L A, Schmidt J E, Van Wier S, Chang X B, Bjorklund C C, Fonseca R and Bergsagel P L *et al* 2011 Cereblon expression is required for the antimyeloma activity of lenalidomide and pomalidomide *Blood* **118** 4771–9

[44] Lu G, Middleton R E, Sun H, Naniong M, Ott C J, Mitsiades C S, Wong K K, Bradner J E and Kaelin W G Jr 2014 The myeloma drug lenalidomide promotes the cereblon-dependent destruction of Ikaros proteins *Science* **343** 305–9

[45] Kronke J, Udeshi N D, Narla A, Grauman P, Hurst S N, McConkey M, Svinkina T, Heckl D, Comer E and Li X *et al* 2014 Lenalidomide causes selective degradation of IKZF1 and IKZF3 in multiple myeloma cells *Science* **343** 301–5

[46] Gandhi A K, Kang J, Havens C G, Conklin T, Ning Y, Wu L, Ito T, Ando H, Waldman M F and Thakurta A *et al* 2014 Immunomodulatory agents lenalidomide and pomalidomide co-stimulate T cells by inducing degradation of T cell repressors Ikaros and Aiolos via modulation of the E3 ubiquitin ligase complex CRL4(CRBN) *Br. J. Haematol.* **164** 811–21

[47] Kronke J, Fink E C, Hollenbach P W, MacBeth K J, Hurst S N, Udeshi N D, Chamberlain P P, Mani D R, Man H W and Gandhi A K *et al* 2015 Lenalidomide induces ubiquitination and degradation of CK1alpha in del(5q) MDS *Nature* **523** 183–8

[48] Ito T and Handa H 2015 Myeloid disease: Another action of a thalidomide derivative *Nature* **523** 167–8

[49] Matyskiela M E, Lu G, Ito T, Pagarigan B, Lu C C, Miller K, Fang W, Wang N Y, Nguyen D and Houston J *et al* 2016 A novel cereblon modulator recruits GSPT1 to the CRL4 (CRBN) ubiquitin ligase *Nature* **535** 252–7

[50] Hansen J D, Condroski K, Correa M, Muller G, Man H W, Ruchelman A, Zhang W, Vocanson F, Crea T and Liu W *et al* 2017 Protein degradation via CRL4(CRBN) ubiquitin ligase: discovery and structure-activity relationships of novel glutarimide analogs that promote degradation of aiolos and/or GSPT1 *J. Med. Chem.*

[51] Hagner P R, Man H W, Fontanillo C, Wang M, Couto S, Breider M, Bjorklund C, Havens C G, Lu G and Rychak E *et al* 2015 CC-122, a pleiotropic pathway modifier, mimics an interferon response and has antitumor activity in DLBCL *Blood* **126** 779–89

[52] Matyskiela M E, Zhang W, Man H W, Muller G, Khambatta G, Baculi F, Hickman M, LeBrun L, Pagarigan B and Carmel G *et al* 2018 A cereblon modulator (CC-220) with improved degradation of ikaros and aiolos *J. Med. Chem.* **61** 492–503

[53] Winter G E, Buckley D L, Paulk J, Roberts J M, Souza A,, Dhe-Paganon S and Bradner J E 2015 Phthalimide conjugation as a strategy for *in vivo* target protein degradation *Science* **348** 1376–81

[54] Sakamoto K M, Kim K B, Kumagai A,, Mercurio F, Crews C M and Deshaies R J 2001 Protacs: chimeric molecules that target proteins to the Skp1-Cullin-F box complex for ubiquitination and degradation *Proc. Natl. Acad. Sci. USA* **98** 8554–59

[55] Lai A C, Toure M, Hellerschmied D, Salami J, Jaime-Figueroa S, Ko E, Hines J and Crews C M 2016 Modular PROTAC design for the degradation of oncogenic BCR-ABL *Angew. Chem. Int. Ed. Engl.* **55** 807–10

[56] Winter G E, Mayer A, Buckley D L, Erb M A, Roderick J E, Vittori S, Reyes J M, di Iulio J, Souza A and Ott C J *et al* 2017 Bromodomain proteins function as master transcription elongation factors independent of CDK9 recruitment *Mol. Cell* **67** 5–18 e19

Chapter 4

Fluorescent magnetic beads for medical diagnostics

Satoshi Sakamoto, Mamoru Hatakeyama and Hiroshi Handa

4.1 Introduction

Science and technology have evolved in accordance with the development of novel materials and tools [1]. A unique technology may open the way for a new research field. In the biomedicine field, an appropriate use of functional materials that have a specific function enables efficient diagnosis and treatment [2]. One of the improvement points desired in a medical field includes the shortening of time for taking a disease diagnosis and body fluid test. If short-time diagnosis of diseases is achieved with absolute accuracy, medical doctors will be able to develop a therapeutic principle flexibly and provide precise treatment for patients. In addition, it will be possible for patients to improve QOL (quality of life) physically and mentally. When diseased sites are excised during the operation of a breast cancer patient, a cancer metastasis to lymph node adjacent to the diseased site is checked. A pathologist closely observes an ultrathin section sample of the excised lymph nodes as a tissue specimen under a light microscope and judges cautiously if there is cancer metastasis in the lymph node. Commonly, a diagnostic outcome is received over at least 30 min after removal of a lymph node. During the operation, the patient is forced to stand ready on an operating table. If the sample is positive for a lymph node metathesis, the major surgery that dissects all lymph nodes around the diseased sites is required. If the sample is negative, dissection surgery will not be conducted. Thus, rapid diagnosis of cancer metastasis leads to a burden reduction to patients and development of a novel disease diagnostic method with high sensitivity and precision is eagerly anticipated.

Nowadays, blood drawing and/or urine collection is conducted before consultations by a medical doctor and patients often wait for medical examination over several tens of minutes until test results are available. When the test is outsourced, the test results will be obtained after several days. If a definitive diagnosis of the

doi:10.1088/978-0-7503-1584-5ch4

disease is done during a physical examination, the medical doctor can treat appropriately and the QOL of patients will be improved. Currently, a consultation time per one patient is just several minutes, at most 10 min. It is more likely that the consultation time becomes less than several minutes, but it is less likely that the consultation time becomes more than 10 min. Thus, lots of people desire a rapid disease diagnosis utilizing the cutting-edge technology.

Several biomarkers provide much information on disease progression and prognosis [3–6], and have been widely used in the diagnosis of disease and at the point of care. Various bioassays, including immunoassay, are used in the measurement of biomarkers in both research and clinical areas [7, 8]. However, most bioassays often take a long time to obtain results. Conventional immunoassays often show low signals due to the use of an enzyme-modified antibody. Various approaches have been examined to solve the intrinsic issues of sensitivity, and utilization of functional materials is one of the solutions for these issues. Magnetic particles are attractive functional materials because magnetic force frequently improves efficiency and sensitivity [9]. Although several highly sensitive immunoassays that simply involve magnetic collection of functionalized magnetic particles have been developed, they still take several hours to obtain assay results [10, 11].

To meet such needs in the medical field, we have addressed the development of a rapid diagnosis method with high precision and sensitivity. We developed unique polymer-coated magnetic beads with several ferrite nanoparticles [12, 14]. Here, we conceived that shortening of a diagnosis time for diseases is achieved by using novel fluorescent magnetic beads that add fluorescence to the polymer-coated magnetic beads. An ELISA (enzyme-linked immunosorbent assay) method based on an antigen–antibody reaction that takes more than one hour has been widely used as a standard test in a normal diagnosis [15, 16]. If antibody-labeled fluorescent magnetic beads can be collected or concentrated magnetically on an antigen-coated plate, the antigen–antibody reaction will be accelerated locally on the plate because of the concentration of both antigen and antibody. Then, target antigens will be detected rapidly by direct fluorescence measurement of the fluorescent magnetic beads collected on the plate, just after the antigen–antibody reaction. In conventional immunoassay, the amplification step that obtains the assay signal by an enzyme reaction after the antigen–antibody reaction is required to detect the target molecules. This assay process does not involve amplification reactions such as enzymatic reaction or PCR reaction. Thus, we thought that the application of fluorescent magnetic beads would lead to a dramatic shortening of the immunoassay time.

4.2 Production of novel fluorescent magnetic beads (fluorescent ferrite beads; FF beads)

For the purpose of efficient isolation and purification of target proteins for bioactive substances such as drugs, we have developed dispersive polymer-coated magnetic beads with a size of around 200 nm in diameter [12, 13]. Core magnetic particles of the polymer-coated magnetic beads are ferrite (magnetite), the size of which is 30–40 nm. The polymer composite of the beads is dual. Copolymer composed of styrene and GMA (glycidyl methacrylate) is covered with several ferrite particles and poly-GMA is covered with the copolymer. We named the polymer-coated magnetic beads FG beads after *ferrite* and *GMA* (figure 4.1). FG beads can be collected easily by a permanent magnet. One of the notable features that FG beads possess includes extremely low binding to irrelative proteins and we believe that this feature brings a significant improvement to the S/N ratio in bioassay using FG beads. In addition, we found the noteworthy feature that FG beads disperse well in organic solvents without damaging the shape of the beads and thought that new functional beads that add FG beads to fluorescence taking advantage of organic solvents will be able to be utilized as a practical bioassay probe.

We selected a hydrophobic fluorescent Eu complex ($Eu(TTA)_3(TOPO)_2$) as the fluorophore that introduces FG beads [17–20]. $Eu(TTA)_3(TOPO)_2$ shows a large Stokes shift (Ex: 340 nm; Em: 618 nm) and a long fluorescence lifetime (over ms). Therefore, the material including $Eu(TTA)_3(TOPO)_2$ can be measured by time-resolved fluorometry with a much higher S/N ratio. When we attempted encapsulation of $Eu(TTA)_3(TOPO)_2$ into the polymer layer of FG beads using hydrophilic organic solvents, new beads that bear about 1.0×10^6 molecules (at a maximum) of $Eu(TTA)_3(TOPO)_2$ per one bead was obtained. We named the new fluorescent magnetic beads FF beads after *fluorescence* and *ferrite* (figures 4.2(a) and (b)) [21].

In order to confirm whether FF beads can be used effectively as an immunoassay probe, we evaluated the physical properties of FF beads and found that they possessed the following features. (1) FF beads can be collected easily by a permanent magnet. (2) Eu complexes are not on the surface of FF beads, but are retained firmly inside the polymer layer of the beads. (3) Eu complexes in FF beads do not leak out to water during both magnetic collection and redispersion. (4) FF beads possess the same size and shape as FG beads and show a high dispersibility in water. (5) A high fluorescence intensity of FF beads enables the measurement of a tiny amount of the sample. (6) A single FF bead contains a lot of Eu complexes and can be observed under a fluorescent microscope (figure 4.2(c)). Even if the amount of Eu complexes in FF beads increases, concentration quenching does not happen. (7) FF beads that encapsulate $Eu(TTA)_3(TOPO)_2$ having a long fluorescence lifetime show low background noise in the time-resolved fluorometry. (8) Fluorescence emission of FF

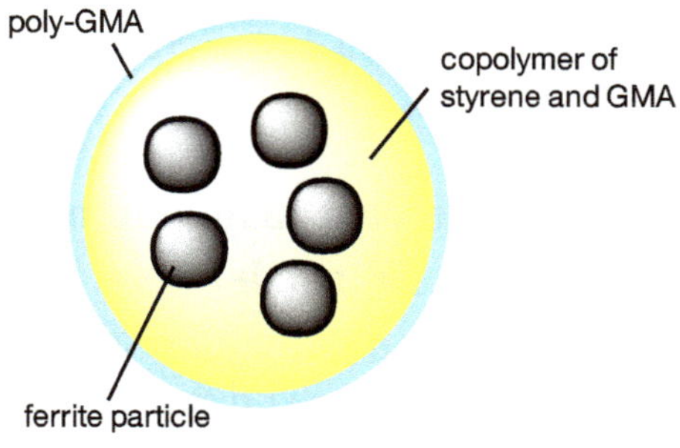

Figure 4.1. Structure of polymer-coated magnetic beads (FG beads).

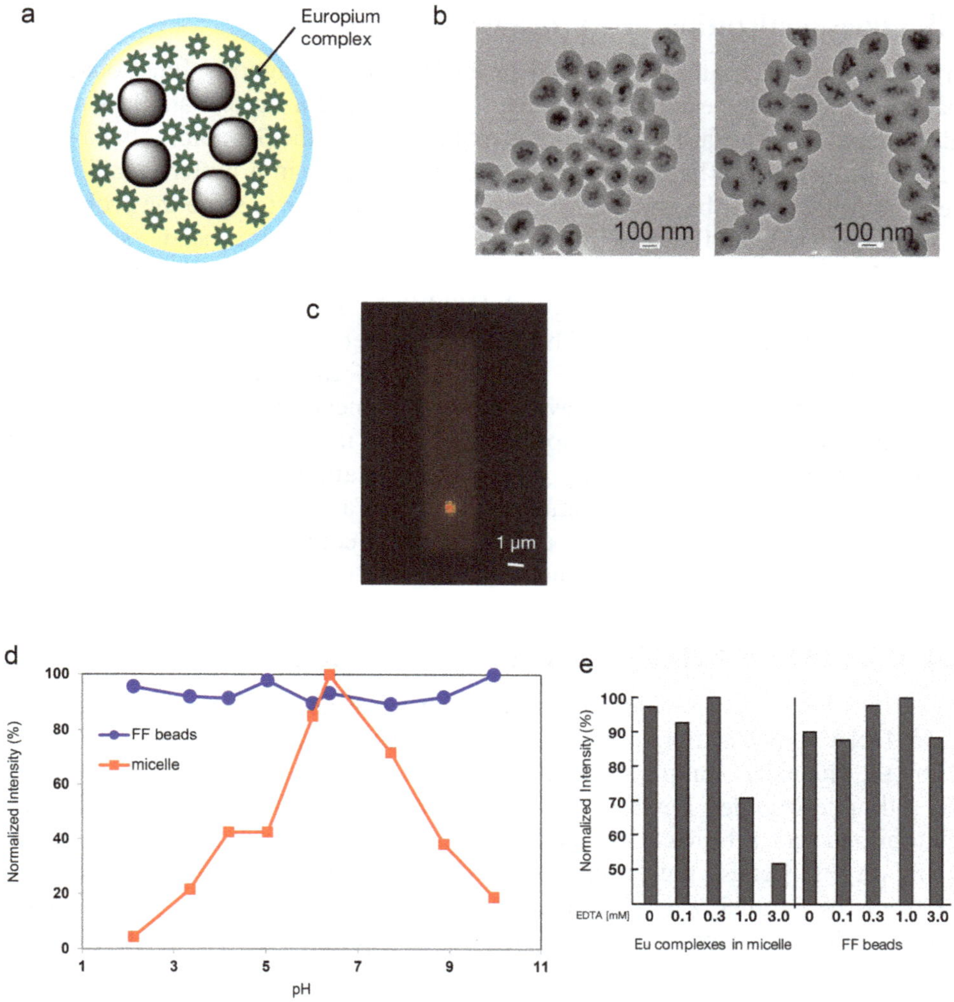

Figure 4.2. (a) Structure of fluorescent magnetic beads (FF beads). (b) TEM images of FG beads (left) and FF beads (right). (c) Observation of a single FF bead under fluorescent microscopy. (d) Stability of the fluorescence intensity of FF beads in a broad range of pH; (e) stability of the fluorescence intensity of FF beads under the presence of EDTA.

beads is stable with a broad range of pH and is not canceled under the presence of EDTA (ethylenediaminetetraacetic acid) known as a fluorescence quencher (figures 4.2(d) and (e)). (9) FF beads are quite stable and can be stored for at least six months without any damage and loss of fluorescence intensity. (10) As with FG beads, FF beads also have the ability of extremely low binding to irrelative proteins. Judging from these features, we believe that FF beads would be a practical probe capable of quantitative measurement of biomarkers with high precision and sensitivity in disease diagnosis.

4.3 Functionalization of FF beads

In order to use FF beads as a bioassay probe for disease diagnosis, functionalization that immobilizes a specific biomolecule (a protein and a nucleic acid) as a ligand adequately on the surface of the beads is required. Here, the possibility that a specific interaction between a ligand on FF beads and its target in a buffer is inhibited by steric hindrance must be taken into consideration. In order to solve the problem, we thought that a ligand should be set at a distance from the beads' surface and immobilization of a ligand through a linker (a spacer) on the beads should be examined. As with the case of FG beads [12–14], EGDE (ethylene glycol diglycidyl ether) was introduced into the FF beads as the linker. Since the ends of EGDE are epoxide, various ligands can be immobilized on the beads through chemical transformation of epoxide. However, the requirement of the linker depends on the kinds of ligands. The linker is not required in the immobilization of DNA on the beads. In the immobilization of a protein such as an antigen and an antibody on the beads a condensation agent is commonly used. As an alternative, a His-tag fused protein can be immobilized on the beads without the condensation agents by making use of the epoxide ring opening reaction.

4.4 Rapid DNA hybridization using magnetic collection of FF beads

To verify whether a rapid bioassay using FF beads is achievable, we first examined rapid DNA hybridization using magnetic collection of DNA-coated FF beads. Using the chemical bonding through the epoxide ring opening by amine of guanine (G) a double-stranded DNA with a single-stranded G-rich terminal sequence was immobilized on FF beads. Through alkali heat denaturation of the double-stranded DNA on FF beads, the specific single-stranded DNA-coated FF beads were prepared. In parallel, a complementary or a non-complementary single-stranded DNA for the sequences of the single-stranded DNA immobilized on FF beads was coated on a glass slide. Then, the DNA hybridization system in which the single-stranded DNA-coated FF beads are collected by a permanent magnet on the single-stranded DNA-coated glass slide was attempted. DNA hybridization was evaluated by fluorescence observation of the FF beads left on the glass slide (figure 4.3(a)). In the case that magnetic collection of FF beads was not applied, it took over 20 h until DNA hybridization of each single-stranded complementary DNA was completed. On the other hand, when magnetic collection of the single-stranded DNA-coated FF beads using a permanent magnet on the slide was conducted, DNA hybridization was completed in only 1 min (figure 4.3(b)). In the cases of using the non-complementary single-stranded DNA, DNA hybridization did not occur and fluorescence derived from FF beads was not detected (figure 4.3(c)). These results indicate that DNA hybridization between DNAs having complementary sequences can be promoted magnetically.

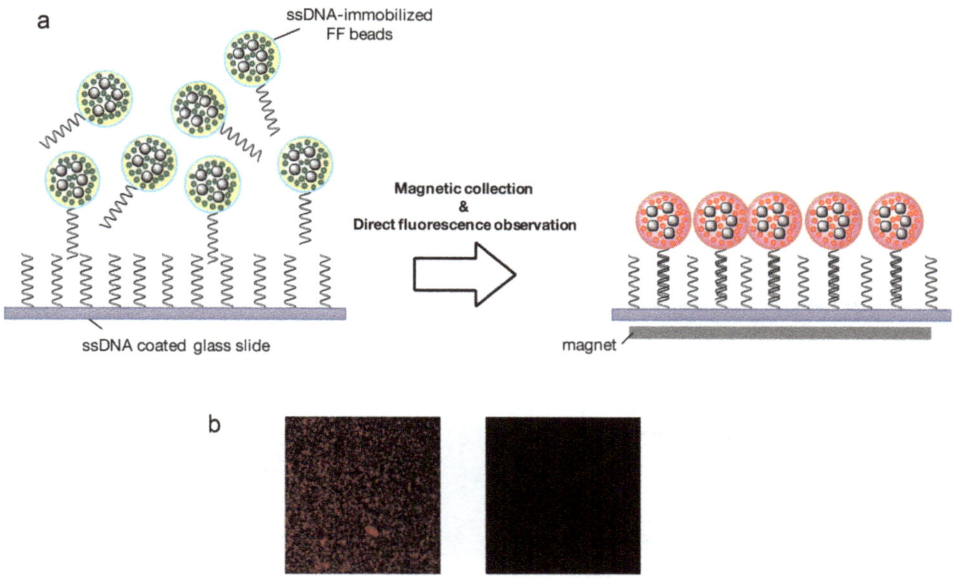

Figure 4.3. (a) Scheme of a rapid hybridization using DNA-immobilized FF beads. (b) Fluorescent observation of the glass slide by a fluorescent microscope after 1 min magnetic collection of DNA-immobilized FF beads. Left: the results when using complementary DNA. Right: the results when using non-complementary DNA.

4.5 Rapid sandwich immunoassay introducing magnetic collection of FF beads: fast detection of the heart failure biomarker BNP

BNP (brain natriuretic peptide) is a hormone secreted by the heart and a representative disease marker of heart failure [22, 23]. The BNP concentration of heart failure patients increases depending on severity. The BNP concentration in the blood plasma of healthy persons is usually less than 20 pg mL^{-1}. Concentrations of 20–40 pg mL^{-1} BNP are associated with heart disease and those of over 100 pg mL^{-1} BNP are associated with heart failure [24, 25]. We obtained the two high-quality anti-BNP antibodies and the purified BNP from a pharmaceutical company and addressed the construction of a rapid sandwich immunoassay with FF beads using BNP materials. One of the anti-BNP antibodies (the capture antibody) was immobilized on a plate and another anti-BNP antibody (the detection antibody) was immobilized on FF beads. Then, we examined the detection of BNP by the sandwich immunoassay introducing magnetic collection of antibody-coated FF beads. Firstly, the antigen–antibody reaction between the purified BNP and anti-BNP antibody on the plate was performed. The capture of BNP by the anti-BNP antibody on the plate was set at 2 min. Next, anti-BNP antibody-coated FF beads were collected magnetically on the plate using a permanent magnet and a second antigen–antibody reaction between the purified BNP and anti-BNP antibody on the FF beads was performed. The magnetic collection of FF beads in the antigen–antibody reaction was set from 0.5 to 3 min. Finally, unbound BNP and anti-BNP

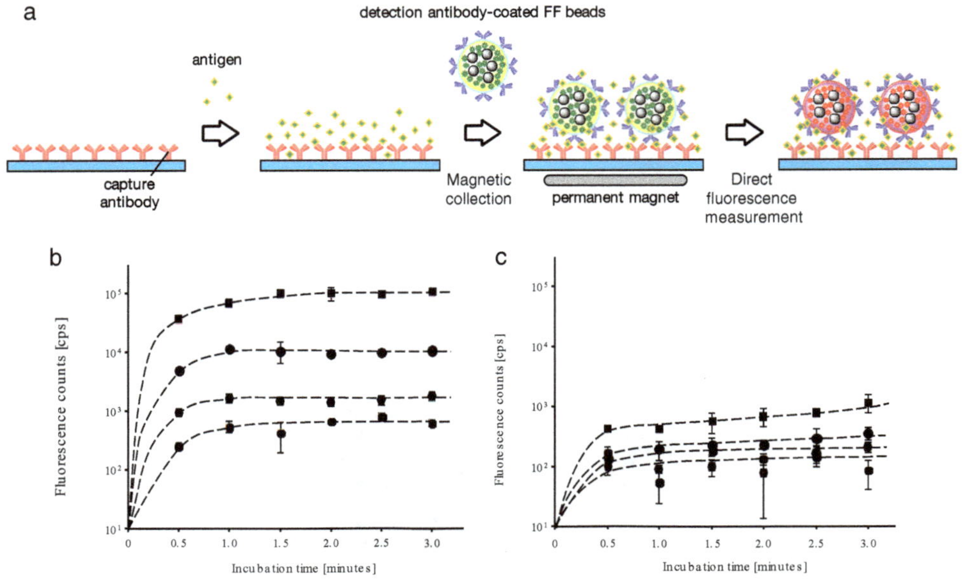

Figure 4.4. (a) Scheme of the magnetically prompted sandwich immunoassay introducing a magnetic collection of antibody-coated FF beads; (b) and (c) results of BNP detection with magnet (b) and without magnet (c). Square, circle, hexangular, and rounded rectangle represent 200, 20, 2.0, and 0 pg mL^{-1} of BNP, respectively. All data are expressed as mean (SD) of three measurements.

antibody-coated FF beads were removed by a wash step and fluorescence measurement was conducted directly. The wash step was set at 1.5 min. The total assay time from the addition of the sample (BNP) to the fluorescence measurement was about 7 min at a maximum (figure 4.4(a)). As a result, BNP was able to be detected by only half a minute of magnetic collection of FF beads and the detection signal of BNP reached the maximum with 1.5 min magnetic collection of FF beads (figure 4.4(b)). In addition, the dose-dependent BNP detection was observed. When the magnetic collection of FF beads was not applied to the sandwich immunoassay using FF beads, the detection signal of BNP was highly variable and did not reach the maximum in a short time (figure 4.4(c)). Accordingly, the antigen–antibody reaction time by the introduction of the magnetic collection of antibody-coated FF beads to the sandwich immunoassay system was reduced to less than one-tenth compared with the conventional immunoassay. Then, we found that only 2.0 pg mL^{-1} of BNP was able to be detected with high accuracy within 5 min after addition of the sample [26].

4.6 Rapid sandwich immunoassay introducing magnetic collection of FF beads: fast detection of the prostate cancer biomarker PSA

PSA is a 33–34 kDa glycoprotein enzyme and is a member of the peptidase family [27]. Although the detailed functions of PSA remain unclear, it is thought that PSA assists sperm movement by modulating the semen viscosity. PSA has been widely used a diagnostic biomarker for prostate cancer and the PSA content in serum increases in prostate cancer patients. However, since the PSA content also increases in prostate inflammation and prostate enlargement, a diagnosis of prostate cancer based on PSA content should be made cautiously. From previously reported data, healthy individuals commonly have a low concentration of PSA (<0.1 ng mL^{-1}) and a PSA content of less than 4.0 ng mL^{-1} is a normal level. The probability of prostate cancer is estimated to be around 30% when the PSA content ranges from 4.0 to 10 ng mL^{-1}. Hence, clinical examination of prostate cancer requires detection of PSA in serum ranging from 0.1 to 10 ng mL^{-1} [28, 29]. In light of this information, we examined the detection of PSA by the rapid sandwich immunoassay system introducing magnetic collection of anti-PSA antibody coated FF beads. We employed two proven anti-PSA antibodies that do not show cross reactivity and pure PSA. The anti-PSA antibody recommended as the capture antibody was coated on a plate and another anti-PSA antibody recommended as the detection antibody was coated on FF beads. As with the case of BNP, the antigen–antibody reaction between PSA and anti-PSA antibody on the plate was performed in the first step of the sandwich immunoassay. The capture of PSA by the anti-PSA antibody on the plate was set at 3 min. Subsequently, the anti-BNP antibody-coated FF beads were collected magnetically on the plate using a permanent magnet and a second antigen–antibody reaction between PSA and anti-PSA antibody on FF beads was performed. The magnetic collection of FF beads in the antigen–antibody reaction was set from 1 to 30 min. Then, unbound PSA and anti-PSA antibody-coated FF beads were removed by a wash step and fluorescence measurement was conducted directly. The wash step was set at 1 min. As a result, PSA was able to be detected by only 1 min magnetic collection of FF beads (figure 4.5(a)). Meanwhile, unlike the case of BNP, the detection signal of PSA reached the maximum with over 10 min magnetic collection of FF beads. A dose-dependent PSA detection was also observed. Interestingly, in all cases where magnetic collection of FF beads was not used, the

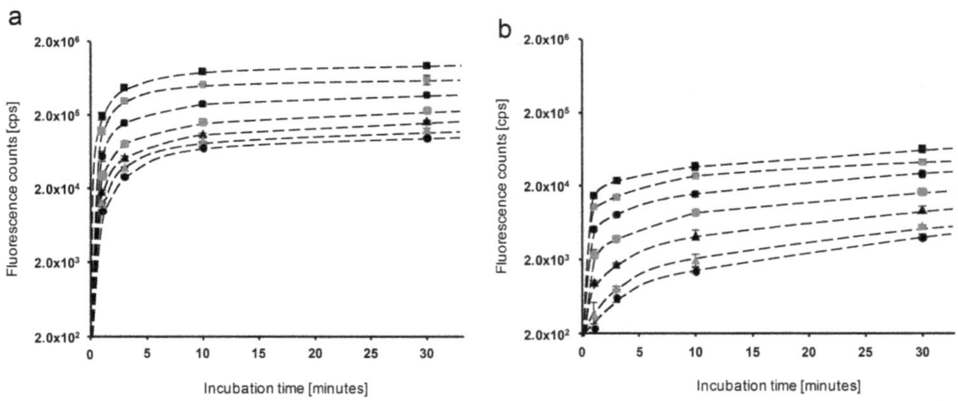

Figure 4.5. Results of PSA detection with a magnet (a) and without a magnet (b). Black square, gray square, black rounded rectangle, gray rounded rectangle, black triangle, gray triangle, and black circle represent 6.3, 2.0, 0.64, 0.20, 0.064, 0.020, and 0 ng mL^{-1} of PSA, respectively. All data are expressed as mean (SD) of three measurements.

PSA detection signal was less stable than the cases with magnetic collection of FF beads (figure 4.5(b)). Finally, we successfully detected 0.02 ng mL^{-1} of PSA within 5 min after addition of the sample [26].

On receiving these results, we investigated the performance of the rapid sandwich immunoassay system using antibody-coated FF beads by comparing the conventional immunoassay. Using PSA samples, the concentrations of which were already known, we measured PSA in serum by the rapid sandwich immunoassay system introducing magnetic collection of anti-PSA antibody-coated FF beads. As a result, a good correlation was obtained between PSA concentrations measured by the rapid sandwich immunoassay and PSA concentrations measured by the conventional immunoassay [26]. Including the results of BNP, these results clearly indicate that the magnetic force enhances and accelerates specific interactions between antigens and antibodies, and the application of magnetic collection of FF beads in a particular space to the immunoreaction can dramatically improve both the total assay time and the number of steps required. Thus, we were able to demonstrate that the rapid sandwich immunoassay system introducing magnetic collection of FF beads is a valid and effective immunoassay system.

4.7 Rapid immunostaining introducing magnetic collection of FF beads

Following the success of the rapid sandwich immunoassay using antibody-coated FF beads, we next examined whether a rapid immunoreaction system using FF beads can be applied to a tissue diagnosis such as an intraoperative pathological diagnosis. In the tissue diagnosis, there is a pathological diagnosis that detects cancer cells in a tissue under the microscope and an immunostaining that detects cancer cells in a tissue by use of an antibody to a specific antigen. The intraoperative pathological diagnosis is used in the diagnosis of cancer metastasis to a primary lymph node (sentinel lymph node) that cancer cells metastasize from primary lesions. However, identification of cancer cells with an abnormal morphology in the tissue by the rapid pathological diagnosis requires much skill. There is a possibility that cancer cells are missed in the case of a few cancer cells or by preparation of a tissue specimen. Staining of the sentinel lymph node using pigments has been used for intraoperative diagnosis. However, a pathological diagnosis with pigments is time-consuming and often leads to misdiagnosis as a result of human error. Therefore, immunostaining has been used to detect cancer cells with high accuracy and reproducibility. To stain only specific cancer cells, an antibody that does not recognize a normal lymph node but recognizes metastatic cancer cells (for instance, the antibody to cytokeratin that organizes the cytoskeleton of cancer cells) is used. As conventional immunostaining takes several hours (and sometimes more) to obtain a definitive diagnosis, we expected that immunostaining introducing magnetic collection of antibody-coated FF beads would be completed in a shorter time.

We selected EGFR (epidermal growth factor receptor) as a target marker which identifies cancer cells and xenograft samples of A431 human epidermoid carcinoma cell, in which EGFR is highly expressed, as representative cancer cells. EGFR is

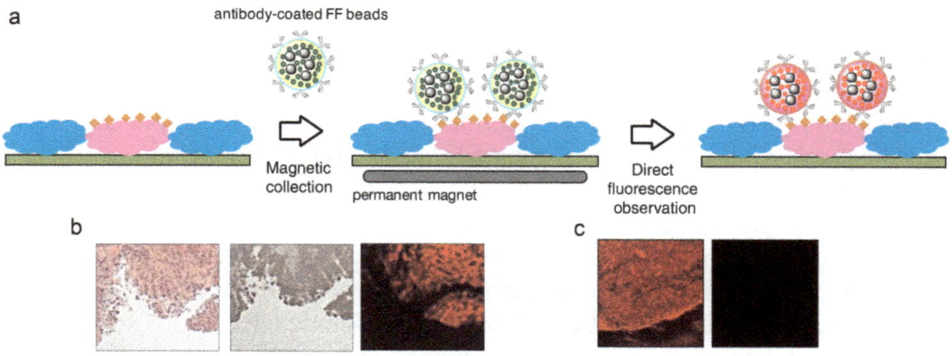

Figure 4.6. (a) Scheme of magnetically prompted immunohistochemical staining introducing antibody-coated FF beads. (b) Immunohistochemical staining of A431 cells (human epidermoid cancer cell) using HE staining (left), polymer-embedded staining (middle), and anti-EGFR antibody-coated FF beads (right). (c) Immunostaining of ultrathin sections using antibody-coated FF beads. Left: The results when using antibody-coated FF beads. Right: The results when using antibody-uncoated FF beads.

known as a key disease marker to understand malignancy because it plays an important role in cancer cell survival and proliferation [30, 31]. As with the rapid sandwich immunoassay, anti-EGFR antibody-coated FF beads were prepared through chemical immobilization of anti-EGFR antibody on FF beads. Then, the anti-EGFR antibody-coated FF beads were treated with fixed samples of cancer cells. In this step, a permanent magnet was attached beneath the fixed samples to promote the immunoreaction between antigens expressed on the surface of carcinoma cells and antibodies immobilized on FF beads. The magnetically promoted immunoreaction time was set at 10 min. After the magnet was separated from the fixed samples, the samples were washed to remove unbound anti-EGFR antibody-coated FF beads and were observed directly by fluorescence microscopy (figure 4.6(a)). As a result, red fluorescence derived from the FF beads was observed in the samples treated with anti-EGFR antibody-coated FF beads. The areas of fluorescence corresponded to those identified by the conventional polymer-labeled method. By contrast, red fluorescence derived from FF beads was hardly observed in the samples where the EGFR expression level was low (figure 4.6(b)) [26]. We also examined the rapid immunostaining of ultrathin sections prepared from excised sentinel lymph nodes using anti-cytokeratin antibody-coated FF beads. As is the case in EGFR, cancer cells were successfully identified within 20 min after addition of FF beads on the ultrathin sections (figure 4.6(c)).

4.8 Conclusion

We successfully developed magnetically collectable fluorescent magnetic beads (FF beads) that have high fluorescence intensity. FF beads can be functionalized through chemical immobilization of DNA and antibody on their surface. Functionalized FF beads allow a rapid bioassay with high sensitivity and precision by the use of magnetic collection of FF beads. Especially, the rapid sandwich immunoassay introducing a magnetic collection of antibody-coated FF beads shows superior quantitative performance and high reproducibility. It is hoped that the rapid sandwich immunoassay using FF beads will be put to practical use in the medical field. In addition to immunoassay, we found that FF beads can be applied to immunostaining. By combining the pathological diagnosis, the rapid and secure tissue diagnosis of diseased tissues using FF beads would be possible.

The successful rapid and sensitive detection of target antigens by magnetic collection of FF beads in the immunoreaction step can be explained by the magnetic concentration of both antigens and antibodies locally on the plate or samples, followed by direct fluorescent measurement without signal amplification. In other words, the magnetic concentration of biomolecule-coated FF beads to a specific place has the potential to accelerate the interaction between biomolecules. Several immunoreaction platforms that use magnetic force to improve their efficiency have been reported [32–35]. These mainly involve removal of unbound capture antibodies and there are a few bioassays that accelerate specific biomolecule interactions and shorten assay times using magnetic force [36]. The immunoreaction systems introducing magnetic collection of antibody-coated FF beads utilize magnetic force, not to remove the magnetic beads but to enhance the signal for the detection of biomarkers. The simple operation of collecting antibody-coated FF beads magnetically at a specific location in the immunoreaction enables improvements in both sensitivity and total assay time. Compared to conventional platforms, this unique immunoreaction system using FF beads achieves the rapid detection of biomarkers without the use of enzyme reactions. Although all operations of the immunoassay and immunostaining with FF beads are performed manually at this time, hereafter the immunoassay time will be shortened further by refinement of the wash step and automation of all operations. Recently, we successfully produced multicolored FF beads by encapsulation of fluorophores other than the Eu complex into the polymer layer of FG beads. By linking fluorescence to a biomarker, a rapid investigation of multiple biomarkers using multicolored FF beads would be possible. A bioassay introducing magnetic collection of functionalized FF beads would be a feasible alternative to a conventional bioassay such as an immunoassay and an immunohistochemical staining.

References

[1] Mir M, Ishtiaq S, Rabia S, Khatoon M, Zeb A, Khan G M, Ur Rehman A and Ud Din F 2017 *Nanoscale Res. Lett.* **12** 500

[2] Bilan R, Nabiev I and Sukhanova A 2016 *Chembiochem* **17** 2103–14

[3] Brody E N, Gold L, Lawn R M, Walker J J and Zichi D 2010 *Expert Rev. Mol. Diagn.* **10** 1013–22

[4] Rusling J F, Kumar C V, Gutkind J S and Patel V 2010 *Analyst* **135** 2496–511

[5] Teng P N, Bateman N W, Hood B L and Conrads T P 2010 *J. Proteome Res.* **9** 6091–100

[6] Mascini M and Tombelli S 2008 *Biomarkers* **13** 637–57

[7] Xiao T *et al* 2005 *Mol. Cell Proteomics* **4** 1480–6

[8] Boja E, Hiltke T, Rivers R, Kinsinger C, Rahbar A, Mesri M and Rodriguez H 2011 *J. Proteome Res.* **10** 66–84

[9] Liu R, Liu J, Xie L, Wang M, Luo J and Cai X 2010 *Talanta* **81** 1016–21

[10] Pedrero M, Campuzano S and Pingarrón J M 2012 *Electroanalysis* **24** 470–82
[11] Gijs M A, Lacharme F and Lehmann U 2010 *Chem. Rev.* **110** 1518–63
[12] Nishio K *et al* 2008 *Colloid. Surf. B Biointerf.* **64** 162–9
[13] Sakamoto S, Kabe Y, Hatakeyama M, Yamaguchi Y and Handa H 2009 *Chem. Rec.* **9** 66–85
[14] Sakamoto S, Hatakeyama M, Ito T and Handa H 2012 *Bioorg. Med. Chem.* **20** 1990–2001
[15] Van Weeman B and Schuurs A 1971 *FEBS Lett.* **15** 232–6
[16] Engvall E and Perlmann P 1971 *Immunochemistry* **8** 871–4
[17] Matsuya T, Tashiro S, Hoshino N, Shibata N, Nagasaki Y and Kataoka K 2003 *Anal. Chem.* **75** 6124–32
[18] Hemmilä I, Dakubu S, Mukkala V M, Siitari H and Lövgren T 1984 *Anal. Biochem.* **137** 335–43
[19] Zohar O, Ikeda M, Shinagawa H, Inoue H, Nakamura H, Elbaum D, Alkon D L and Yoshioka T 1998 *Biophys. J.* **74** 82–9
[20] Desbiens J, Bergeron B, Patry M and Ritcey A M 2012 *J.Colloid Interface Sci.* **376** 12–9
[21] Hatakeyama M, Mochizuki Y, Kita Y, Kishi H, Nishio K, Sakamoto S, Abe M and Handa H 2009 *J. Magn. Magn. Mater.* **321** 1364–7
[22] Sudoh T, Kangawa K, Minamino N and Matsuo H 1988 *Nature* **332** 78–81
[23] Sudoh T, Minamino N, Kangawa K and Matsuo H 1988 *Biochem. Biophys. Res. Commun.* **155** 726–32
[24] Nakamura M, Endo H, Nasu M, Arakawa N, Segawa T and Hiramori K 2002 *Heart* **87** 131–5
[25] Cowie M R, Struthers A D, Wood D A, Coats A J, Thompson S G, Poole-Wilson P A and Sutton G C 1997 *Lancet* **350** 1349–53
[26] Sakamoto S *et al* 2014 *Clin. Chem.* **60** 610–20
[27] Stenman U H, Leinonen J, Zhang W M and Finne P 1999 *Semin. Cancer Biol.* **9** 83–93
[28] Finne P, Auvinen A, Määttänen L, Tammela T L, Ruutu M, Juusela H, Martikainen P, Hakama M and Stenman U H 2008 *Eur. Urol.* **54** 362–70
[29] Chatterjee S K and Zetter B R 2005 *Future Oncol.* **1** 37–50
[30] De Luca A, Carotenuto A, Rachiglio A, Gallo M, Maiello M R, Aldinucci D, Pinto A and Normanno N 2008 *J. Cell Physiol.* **214** 559–67
[31] Ozawa S, Ueda M, Ando N, Abe O and Shimizu N 1987 *Int. J. Cancer* **39** 333–7
[32] Liu R, Liu J, Xie L, Wang M, Luo J and Cai X 2010 *Talanta* **81** 1016–21
[33] Shinkai M, Wang J, Kamihira M, Iwata M, Honda H and Kobayashi T 1992 *J. Ferment. Bioeng.* **73** 166–8
[34] Nishizono I, Iida S, Suzuki N, Kawada H, Murakami H, Ashihara Y and Okada M 1991 *Clin. Chem.* **37** 1639–44
[35] Liabakk N B, Nustad K and Espevik T 1990 *J. Immunol. Methods* **134** 253–9
[36] Mulvaney S P, Mattoussi H M and Whitman L J 2004 *Biotechniques* **36** 608–9

Chapter 5

Development of surface-modified magnetic nanoparticles for medical applications

Masaaki Kawano, Mamoru Hatakeyama, Masanori Matsui and Hiroshi Handa

5.1 Medical uses of magnetic materials

Due to their magnetic properties, magnetic materials have been used for many years as storage media, sensors, and shields in the field of electronics. By contrast, magnetic materials have played a more limited role in the life sciences and the medical field to date. The magnetic materials that are of particular interest in the biomedical field are the superparamagnetic iron oxide (Fe_3O_4 and γ-Fe2O3, also called magnetite and maghemite, respectively) and ferrite (e.g. $MnFe_2O_4$) nanoparticles that are less than 10 nm in diameter [1–4]. Since magnetic nanoparticles consist of iron and oxygen only, they have excellent biocompatibility and can be administered inside the living body. At present, superparamagnetic magnetic nanoparticles that are coated with biocompatible polymers such as dextran serve as magnetic resonance imaging (MRI) contrast agents, drug carriers, and heating agents in magnetic thermotherapy [5]. However, research that aims to improve all of these applications is ongoing [6]. For example, MRI using magnetic nanoparticles as the contrast agent is limited by poor contrast when exploring organs other than the liver. Superparamagnetic magnetic nanoparticles that carry an immobilized ligand are also widely used in medicine and basic research to separate bioactive substances such as proteins, enzymes and nucleic acids, to screen the blood for tumor cells, to sort cells, and to develop immunoassays (e.g. for detecting infections) [7–15].

5.2 Preparation of magnetic nanoparticles

To prepare magnetic nanoparticles for medical use, naked magnetic nanoparticles are synthesized and allowed to aggregate and precipitate in water. To stabilize the nanoparticles and facilitate their dispersal after administration in the body, the surface of each particle should be coated with a suitable organic or inorganic material. The coating reaction should occur at the time of magnetic nanoparticle

doi:10.1088/978-0-7503-1584-5ch5

synthesis, otherwise the result is large coated aggregates (figure 5.1). Several methods of synthesizing coated magnetic nanoparticles have been reported [16, 17]. A particularly easy method that generates uniform nanoparticles is to synthesize the magnetic nanoparticles in organic solvents while using oleic acid as a capping agent [16]. This generates magnetic nanoparticles that have oleic acid adsorbed to their surface.

For biomedical applications in aqueous environments, this hydrophobic coat should be changed to a hydrophilic coat. One way to achieve this is to use the 'ligand-exchange reaction' (the ligand in this case is the molecule that is adsorbed to the nanoparticle surface). Figure 5.2 depicts a typical ligand-exchange reaction, which involves sonicating oleic acid-coated nanoparticles in the presence of large amounts of the hydrophilic new ligand (e.g. citric acid). However, this procedure has a drawback: since the oleic acid-coated magnetic nanoparticles are hydrophobic, they tend to aggregate in an aqueous solution of citric acid and even sonication does not generate full exchange of the oleic acid molecules with citric acid. Consequently, several other methods that improve the exchange of ligands on oleic acid-coated magnetic nanoparticles have been reported [18–20]. In one of these methods, the oleic acid molecules are exchanged with 2,3-mercaptosuccinic acid [18]. In this method, the ligand-exchange reaction is performed in organic solvents and 2,3-mercaptosuccinic acid is absorbed to the surface of the oleic acid-coated magnetic nanoparticles via its thiol and carboxyl groups. As the ligand-exchange reaction proceeds, the oleic acid molecules break away from the surface of the nanoparticles.

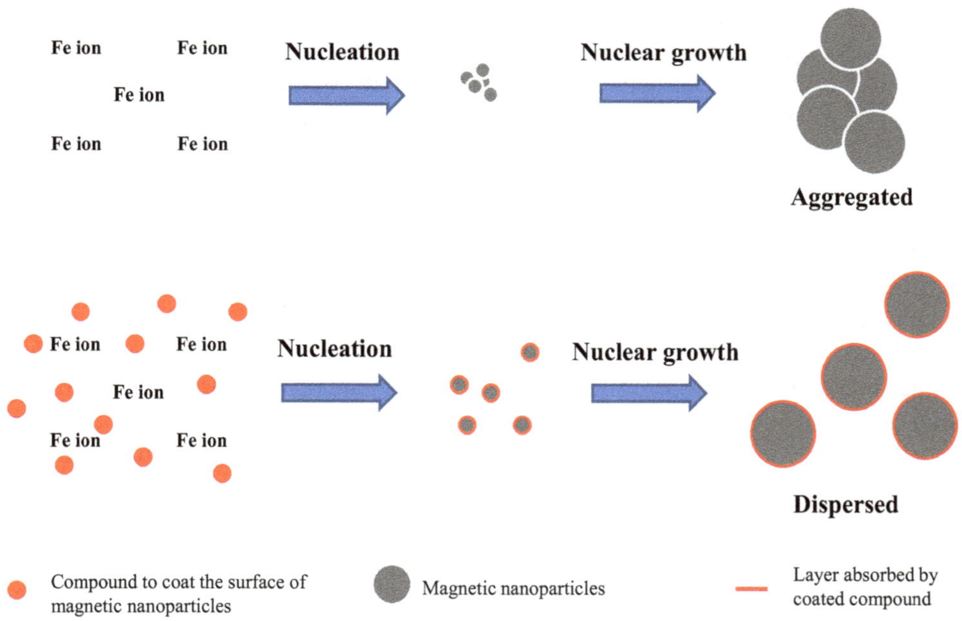

Figure 5.1. Schematic depiction of the process used to synthesize magnetite nanoparticles. If left uncoated, the particles aggregate during their synthesis. However, if the particles are synthesized in the presence of a suitable coating material, they will disperse in the solution.

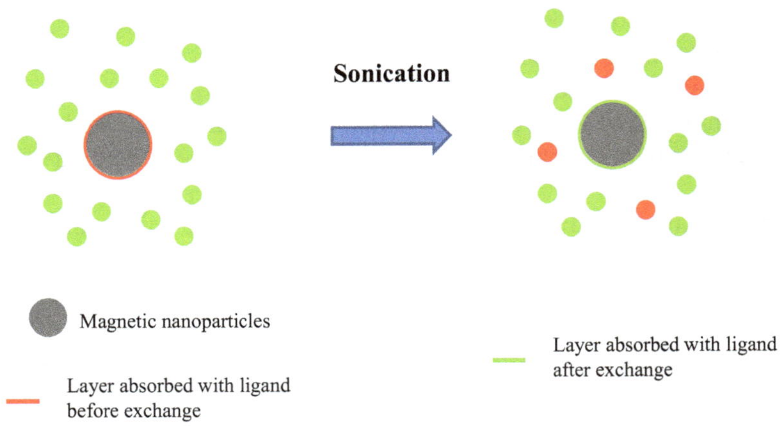

Figure 5.2. Schematic depiction of the typical ligand-exchange reaction that is used to generate magnetic nanoparticles that disperse in hydrophilic solutions.

This causes the dispersibility of the nanoparticles to drop, resulting in their precipitation. When these precipitated particles are placed in water, they again become widely dispersed because of electrostatic repulsion due to the carboxyl groups of the 2,3-mercaptosuccinic acid molecules.

However, although this method results in dispersed nanoparticles with a hydrophilic coat, the two thiol groups of 2,3-mercaptosuccinic acid chelate with the iron (Fe) on the surface of the nanoparticles. This causes the magnetic nanoparticles to disintegrate and decreases their magnetization. Given this limitation, we developed another method to exchange oleic acid with citric acid. This method involves two ligand-exchange reactions: oleic acid is first exchanged with an intermediate compound in an organic solvent, after which that compound is exchanged with citric acid in an aqueous solution. We speculated that compounds with thiol group(s) might be good intermediates because they adsorb well to the surface of magnetite nanoparticles. When we tested a number of potential intermediates (table 5.1) we found that thiomalic acid was an excellent intermediate: the resulting magnetic nanoparticles exhibited very good dispersion stability [21]. Indeed, the resulting nanoparticle diameter in water was almost identical to the diameter of the particles that was measured by transmission electron microscopy (TEM) (figure 5.3).

The thermal decomposition method of synthesizing magnetic nanoparticles makes it possible to separate the nucleation and nuclear growth reactions of the particles. Therefore, it is possible to prepare preparations of magnetic nanoparticles that all have similar specific diameters. Consequently, when we combined the thermal decomposition method with our two-step ligand-exchange method, we were able to obtain monodispersed citric acid-coated magnetic nanoparticles that had particular nanoparticle diameters (4, 8, and 20 nm) (figure 5.3 and table 5.1).

We have shown that all three nanoparticle preparations were useful as T1- and T2-weighted contrast MRI agents (figure 5.4). We also found that when the 20 nm nanoparticle preparation served as an MRI contrast agent in a nude rat model of

Table 5.1. Hydrodynamic diameters of citric acid-coated magnetite nanoparticles that were obtained from oleic acid-coated 4, 8, and 20 nm magnetite nanoparticles by two-step ligand-exchange reactions involving different intermediates.

Primary diameter(nm)	4	8	20
Intermediate		Hydrodynamic diameter (nm)	
(–)	106.4 ± 47.4	414.7 ± 354.1	482.2 ± 207.1
(a)	9.0 ± 2.6	10.1 ± 5.7	19.4 ± 4.7
(b)	5.0 ± 2.7	9.2 ± 5.5	40.3 ± 16.0
(c)	4.2 ± 2.8	8.1 ± 4.2	20.0 ± 6.2
(d)	18.6 ± 2.5	16.3 ± 6.9	68.8 ± 24.4
(e)	12.0 ± 3.2	9.0 ± 4.6	22.6 ± 3.0
(f)	195.7 ± 108.2	199.6 ± 31.8	419.4 ± 168.8

The intermediates that were used are (–) none, (a) thioglycolic acid, (b) thiolactic acid, (c) thiomalic acid, (d) 3-mercaptopropionic acid, (e) 2,3-dimecaptosuccinic acid, and (f) 1,2-ethanedithiol.

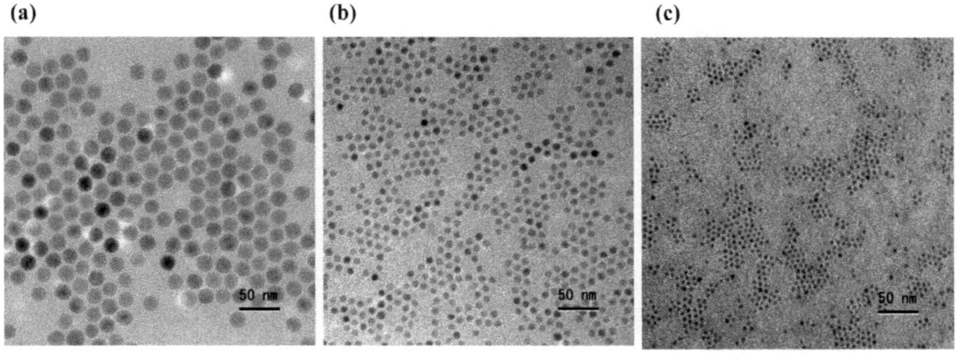

Figure 5.3. Transmission electron microscopy of the magnetite nanoparticles generated by our two-step ligand-exchange reaction. Magnetite nanoparticles of various sizes were synthesized by thermal decomposition in the presence of oleic acid. The oleic acid coat was then exchanged by using thiomalic acid as an intermediate coat, after which the thiomalic acid was exchanged with citric acid. The reaction solvent and the diameter of the resulting particles were (a) 1:4 (v/v) mix of 1-hexadecene:1-octadecene, 4.6 ± 0.5 nm, (b) 1-octadecene, 8.5 ±0.8 nm, and (c) tri-*n*-octylamine, 20.7±1.3 nm.

metastasis to sentinel lymph nodes, it accurately identified a micrometastasis in the cubital lymph node, as confirmed by *ex vivo* image analysis of the extracted lymph node. The secondary (armpit) lymph nodes, which lacked a metastasized tumor, were also negative for tumors by this MRI method (figure 5.5) [22]. Thus, the 20 nm magnetic nanoparticle preparation was useful as an MRI contrast agent for identifying sentinel lymph nodes with metastasis (this study was a collaboration with the Faculty of Medicine of Keio University and the National Institute of Radiological Sciences, Japan). The 20 nm nanoparticles also had magnetic hyperthermia activity (figure 5.6). Thus, our two-step ligand-exchange reaction method generates citric acid-coated magnetic nanoparticles that have specific particle diameters and therefore have unique characteristics that make them suitable for different applications.

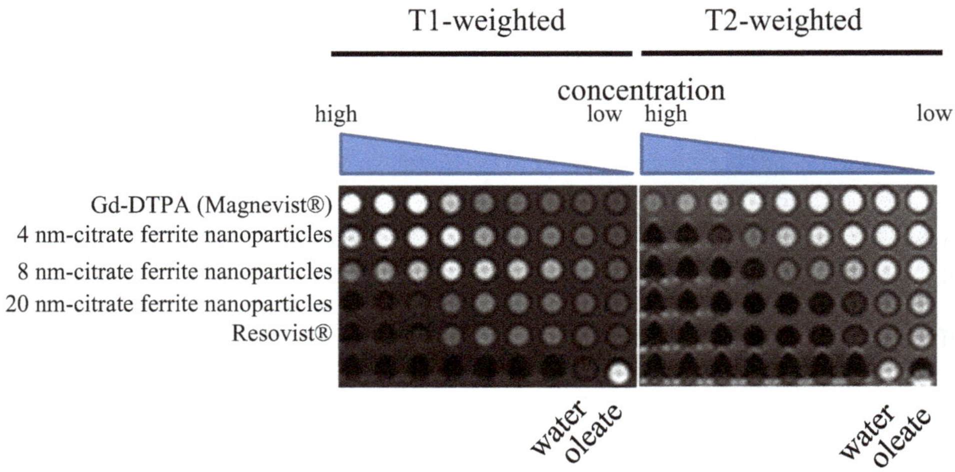

Figure 5.4. The T1- and T2-weighted magnetic resonance images of citric acid-coated 4, 8, and 20 nm magnetite nanoparticles. The metal ion concentrations of each contrast agent were 0.1, 0.075, 0.05, 0.025, 0.01, 0.0075, 0.005, 0.0025, and 0.001 mg mL^{-1}, in descending order.

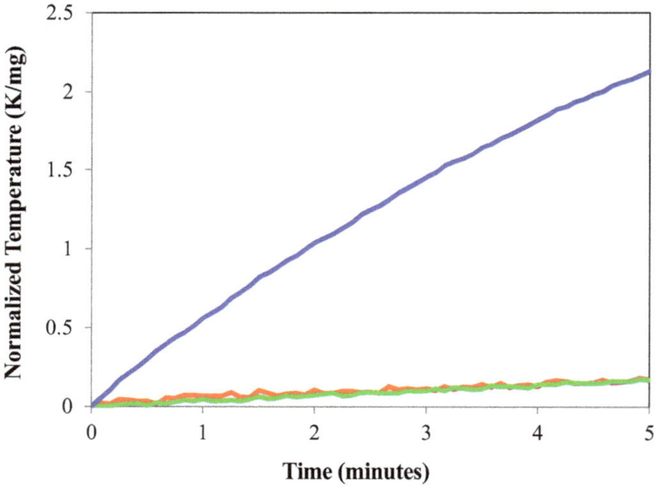

Figure 5.5. Use of citric acid-coated 20 nm magnetite nanoparticles in magnetic resonance imaging to identify a sentinel lymph node bearing a micrometastasis in a rat model of sentinel lymph node metastasis.

5.3 Virus capsid proteins

Depending on the virus, the virus capsid varies in shape and its constituents. For example, the human immunodeficiency virus (HIV) protects its single-stranded RNA genome with two shells composed of a capsid and matrix protein that are finally covered with a double lipid bilayer. In adenovirus (Ad), the linear double-stranded DNA genome is protected by an icosahedral protein shell that consists of hexon and penton proteins.

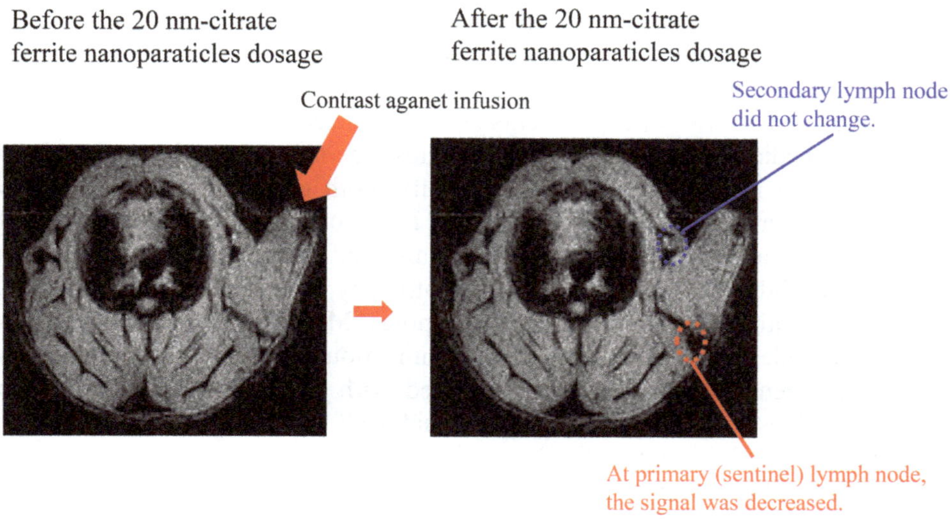

Figure 5.6. Changes in temperature achieved when aqueous dispersions of citric acid-coated magnetite nanoparticles are subjected to an AC electromagnetic field. The red, green, and blue lines indicate the 4, 8, and 20 nm magnetite nanoparticles, respectively.

The virus capsid has several functions. First, it protects the genomic information of the virus from the harsh external environment, thus allowing the virus to survive until it can find a suitable host. Second, it mediates attachment to the target host cells in which it can replicate itself. This function is due to its surface molecules, which were acquired from the preceding host and therefore have highly selective affinity for specific host cells. Third, the virus capsid mediates entry into the host cells by triggering endocytosis. After entry, the virus capsid can enter the cytosol directly. Alternatively, it escapes from the endosome or is trafficked to the center of the cell to the endoplasmic reticulum (ER), from which it then escapes into the cytosol. These cytosol-entry mechanisms involve disrupting the lipid bilayer or crossing the bilayer by using bilayer-residing transporters. In HIV, the enveloped capsid interacts specifically with CD4 and CXCR4/CCR5 on the host cell surface, after which the capsid fuses with the plasma membrane and enters the cytosol directly. In Ad, the capsid is endocytosed and travels in this transport system until the late endosome stage. At that point, it undergoes a change in conformation and releases the Ad protein VI. Ad protein VI associates with the late endosome membrane and ruptures it, thus allowing the capsid to escape into the cytosol. After entering the cytosol, the virus capsid uses various mechanisms to transport its virus genome into the nucleus, where replication will take place. These mechanisms include the release of the virus genome into the cytosol, followed by its attachment to nuclear transport proteins that bring it into the nucleus. Another mechanism is attachment of the capsid itself to the nuclear transport factors. The latter mechanism is employed by the simian virus 40 (SV40): in this case, the transporter is importin. Once inside the nucleus, the capsid immediately releases its genome, thereby triggering the start of genome replication. Hence, a single virus capsid bears various functions due to its constituent proteins. All of these virus capsid protein functions can be used to accurately deliver biological materials to the target cells for medical diagnostic and therapeutic purposes.

We conducted a series of experiments testing the feasibility of using virus capsid proteins for medical purposes. For this, we chose the major capsid protein VP1 of SV40 because it is easy to prepare and modify, both genetically and chemically. SV40 is a member of the polyomavirus family. It was originally isolated in the 1960s from the polio vaccine, in which it was a contaminant [23]. Since the contaminated polio vaccine did not induce any acute cytotoxicity, the SV40 capsid may be a particularly suitable material for medical purposes. Moreover, when polyomavirus virus-like particles (VLPs), which are a non-infectious virus capsid that does not contain viral gene information, are incubated with dendritic cells, they do not activate [24]. This suggests that the SV40 capsid itself is neither toxic nor markedly inflammatory. In addition, the SV40 capsid can be generated from the VP1 protein alone and its configuration is very simple. Specifically, the SV40 capsid consists of 72 pentamers that each consist of five VP1 molecules: thus, each capsid contains 360 VP1 molecules (figure 5.7(A)). The resulting capsid has an icosahedral lattice structure ($T = 7$ dextro) and is approximately 45 nm in diameter [23]. Each VP1 pentamer consists of a ring of five β-barrel-shaped VP1 monomers that are tightly linked by interacting loops between the framework of β-strands. The C-terminal region of each VP1 monomer projects into the neighboring pentamer, thereby tying the pentamers together in the capsid. Several studies show the importance of the C-terminal region of VP1 by examining the effect of mutations in this region: when a mutant VP1 that lacked the C-terminal arm was introduced in insect cells, the VP1 molecules retained their pentameric structure but did not form VLPs [25, 26]. The C-terminal region also contains two calcium ion-binding sites that control the stability of the capsid [27, 28]. Capsid stability is further enhanced by disulfide bonds between cysteine residues in the VP1 pentamers; thus, when calcium is withdrawn in reducing conditions, the C-terminal region triggers capsid disassembly, thereby releasing the encapsulated material [29]. When the calcium-binding sites and the disulfide-bonding cysteine residues are mutated, the mutant VP1s form VLPs normally but exhibit less capsid stability. Their capsid disassembly in the presence of the calcium-chelating agent EGTA and the disulfide-reducing agent DTT is also impaired [27, 29]. Notably, the VP1 pentamers not only regulate capsid stability and disassembly, they also have self-reassembly activity; when the VLPs generated by recombinant polyomavirus VP1 protein in *Escherichia coli* are purified and disassembled into their constituent pentamers by adding EGTA and DTT, the pentamers will reassemble themselves into VLPs *in vitro* when subjected to dialysis with a high salt buffer [28]. Hence, VP1 tightly regulates SV40 capsid assembly, stability, and disassembly. These properties ensure that the encapsulated materials are protected from the harsh environment outside the host cell until capsid entry into the host cell. Subsequently, after traveling to the nucleus, the capsids promptly disassemble into their capsid subunits and release the encapsulated material inside the nucleus.

The VP1 protein also mediates targeted host cell binding of the capsid (one of the natural host cells for SV40 is the African rhesus kidney cell) because it can bind to

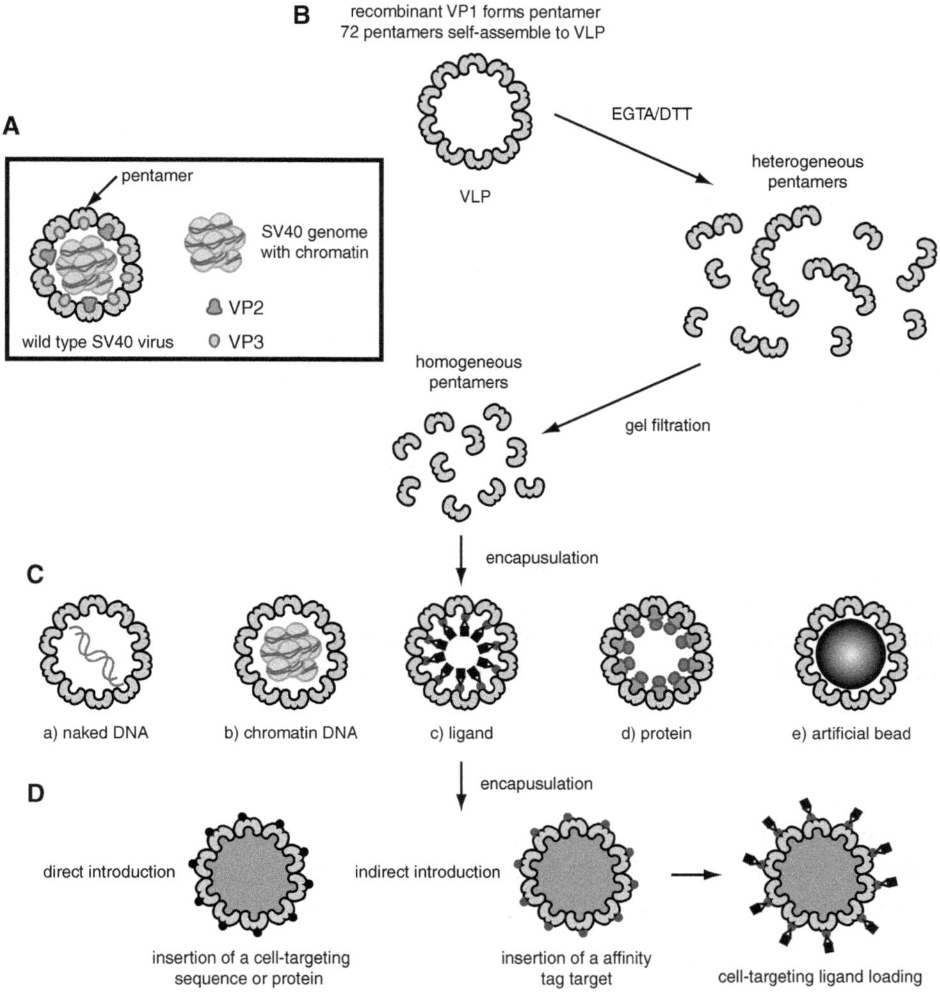

Figure 5.7. Construction of surface-modified nanoparticles that can be used to encapsulate material or load ligands on their surface. (A) Schematic depiction of the structure of the wild-type SV40 virus. The capsid consists of 72 pentamers and VP2 or VP3, which attaches to each pentamer on the inside of the capsid. The core of the capsid contains the double-stranded circular SV40 genome (5243 base pairs), which is arranged with histones. (B) Method of preparing homogeneous samples of monomeric pentamers. The virus-like particles (VLPs) are induced to disassemble by adding EGTA and DTT to the solution, after which gel filtration is performed. (C) Encapsulation of materials within VLPs. (a) Naked DNA or (b) chromatin DNA is encapsulated in nanoparticles during the *in vitro* assembly reaction. The DNA triggers capsid assembly in physiological salt and pH conditions. (c) The affinity tag target is introduced into the interior of the capsid subunit by genetic insertion or chemical cross-linking. The encapsulated materials (including drugs, small chemicals, polymers, peptides, proteins, and artificial beads) are fused to the affinity tag. These materials are then loaded into the capsid during the *in vitro* assembly reaction. (d) Foreign proteins are fused to capsid scaffold proteins such as the C-terminus of VP2. The fusion protein is then loaded during the *in vitro* reaction. Thus, the fusion protein becomes incorporated inside the nanoparticle. (e) Artificial beads made of polystyrene, silica, poly-lactic-*co*-glycolic acid, liposomes, magnetite, gold, or quantum dot particles are coated with pentamers during assembly. (D) Genetic or chemical surface modification of the nanoparticles. In direct introduction (*left*), the appropriate cell-targeting sequence or protein is directly bound to the surface of the nanoparticle. In indirect introduction, the target of a cross-linker, affinity tag, or affinity protein is bound to the surface of the nanoparticle by genetic insertion or chemical cross-linking (*right*). In the latter case, the appropriate cell-targeting ligand is then fused with the cross-linker, affinity tag, or affinity protein. This allows it to be loaded onto the surface of the nanoparticle.

the ganglioside GM1. There are three loops on the capsid surface termed BC, DE, and HI. These loops bind selectively to GM1. After binding to GM1, the capsid enters the host cell. However, since GM1 occurs on various cells, including those that are resistant to SV40 capsid infection, the host cell specificity of SV40 is not fully explained by the ability of its capsid to bind to GM1.

The binding of SV40 capsid to GM1 induces the endocytosis of the capsid. The capsid then travels via the caveosome or endosome to the ER, where it binds to ER membrane proteins and induces membrane transition. This allows the capsid to translocate across the ER membrane into the cytosol [30, 31]. Once in the cytosol, the capsid binds importins via its nuclear localization signal and enters the host cell nucleus. At that point, it disassembles completely, thus releasing the encapsulated genome. Viral genome replication and its gene transcription/translation are then initiated by using the host cell machinery.

The SV40 genome encodes early genes, namely, large T antigen (LTag) and small T antigen (sTag), and late genes, namely, VP1, VP2, and VP3. Thus, LTag and sTag are first transcribed and translated. They promote host cell proliferation and viral genome replication. VP1, VP2, and VP3 are then transcribed and translated: this is critical for viral genome packaging and capsid assembly. The VP1 molecules form VP1 pentamers in the cytosol, VP2 or VP3 associates with these pentamers, and the complexes translocate into the nucleus. In the nucleus, the VP1 pentamers bind together to form the capsid around the 5243 base pairs of the circular double-stranded cellular histone-coordinated DNA that is the SV40 genome (figure 5.7(A)). This capsid formation is driven solely by the self-reassembly activity of VP1: host cell factors are not needed. Thereafter, the SV40 virus induces capsid budding on the plasma membrane or disrupts the plasma membrane: the virus is thus released from the host cell and can seek new host cells.

The minor SV40 capsid protein VP2, or its C-terminally truncated form VP3, is localized inside the capsid. One VP2 or VP3 molecule associates with one VP1 pentamer. Thus, in total, the SV40 capsid bears 72 VP2 or VP3 molecules on its interior surface (figure 5.7(A)). VP2 and VP3 regulate capsid assembly and improve capsid stability and the efficiency of the SV40 genome delivery.

Other polyomaviruses have been found in various species, including mice, cattle, birds, humans, and swine. Their capsid structures are similar to that of the SV40 capsid. They differ from each other in the sequences of their BC, DE, and HI loops in VP1. These differences are responsible for the host cell specificity of these polyomaviruses. For example, these three loops in human BK polyomavirus VP1 interact with GT1b, while in murine polyomavirus VP1, they bind to GD1a.

5.4 Scientific and medical applications of virus capsid protein

As expected, various VLPs can deliver their encapsulated materials accurately and safely to the targeted cell compartment without any loss of their activities. VLPs can be prepared by expressing the recombinant capsid constituents in bacteria, yeast, or insect cells. For example, hepatitis B virus (HBV) VLPs and human papillomavirus (HPV) L1 VLPs have been prepared for use as vaccines by using yeast or insect cells.

In terms of the polyomaviruses, SV40 VLP was first generated in bacteria in 1986 [28]. Since then, various polyomavirus VLPs have been prepared, including murine polyomavirus, human BK polyomavirus, and human JC polyomavirus.

We currently prepare SV40 VLPs by using several steps. First, insect Sf-9 cells are transfected with a recombinant baculovirus that expresses SV40 VP1. The cells are then resuspended and lysed in 20 mM Tris–HCl, pH 7.9 and 1% sodium deoxycholic acid-containing medium that lacks sodium salt. After sonicating the lysates, the intact VLPs are separated and purified by subjecting the lysate twice to 20%–50% (weight per volume) cesium chloride density gradient centrifugation. The fact that SV40 VLPs can be purified with these buffers indicates that SV40 VLPs are stable in both low and high ionic strength buffers as well as in high concentrations of neutral detergent buffers. By contrast, the VLPs derived from enveloped viruses are readily disassembled and form aggregates in both low and high ionic strength buffers and detergent-containing buffers because of the instability of the lipid bilayer on the surface of the VLPs in these buffers; therefore, although the lipid bilayer of enveloped viruses means that their VLPs pass through the cell membrane much more easily than the VLPs from non-enveloped viruses, it complicates their production in the laboratory. Moreover, the VLPs from non-enveloped virus capsids are relatively stable when compared to the VLPs from enveloped viruses. Given that VLPs from non-enveloped viruses are relatively easy to prepare and handle biochemically and genetically, they are particularly suitable for generating medical diagnostics and therapies.

The next step in our laboratory is to dialyze the VLP-containing fractions against a solution that consists of 20 mM Tris–HCl, 100 mM NaCl, and 0.1% Nonidet P-40 [29]. To induce the VLPs to disassemble into pentamers, EGTA and DTT are added to the dialyzed sample. The sample is then fractionated with a gel filtration column to yield the homogenous pentamer sample (figure 5.7(B)).

This sample can then be used to obtain pure VLPs. For this, the homogenous pentamers are dialyzed against certain conditions: the nature of these conditions determines what type of VLP structure will arise. Thus, dialysis against high salt and normal pH conditions consisting of 2 M $(NH_4)_2SO_4$, 2 mM $CaCl_2$, and 20 mM Tris–HCl (pH 7.2) or 1 M NaCl, 2 mM $CaCl_2$, and 20 mM Tris–HCl (pH 7.2) causes the pentamers to reassemble into Triangulation = 1 ($T = 1$), $T = 3$- or $T = 4$-like, or $T = 7$ icosahedral capsids that are 20, 25–35, and 40 nm in diameter, respectively. Under these conditions, the pentamers also sometimes reassemble into tube-like structures that are approximately 40–45 nm in width and have an average length of 120 nm (but the longest exceeds 500 nm) [32]. However, when the VP1 pentamers are dialyzed against physiological salt but acidic pH conditions (150 mM NaCl, 2 mM $CaCl_2$, pH 5.0), they reassemble exclusively into long tube-like structures. By contrast, if the pentamers are dialyzed against physiological salt but basic pH conditions (150 mM NaCl, 2 mM $CaCl_2$, Tris–HCl, pH 9.0), they reassemble into distorted spherical structures. Finally, when they are dialyzed against physiological salt conditions at pH 7, they do not undergo self-reassembly.

The C-terminal region of VP1, which plays a critical role in the interactions between the pentamers in the capsid, also plays an important role in the structures that are adopted by the pentamers in the various dialysis conditions. The dialysis buffer conditions affect the charges of the side chains of the C-terminal region as well as the surface charge of the neighboring pentamer to which the C-terminal region binds. These changes alter the inter-VP1 interactions, thereby generating the icosahedral capsids with varying diameters and tube-like structures. Notably, plant viruses such as tobacco mosaic virus, cowpea chlorotic mottle virus (CCMV; bromovirus group), and bromo mosaic virus (BMV) also yield icosahedral VLPs with various diameters and tube-like structures, as indicated by electron microscopy. Icosahedral capsid structures of any size can be constructed from five-fold and six-fold capsid subunits, and tube-like structures can be constructed from six-fold capsid subunits alone [33]. In the case of SV40, its icosahedral capsids are constructed from pentamers that take two forms (five and six-fold pentamers) depending on the configuration of the inter-pentameric binding through the C-terminal region of VP1 [34]. Therefore, it is theoretically possible to construct icosahedral structures of any size from the pentamer solution alone. As will be described further below, this ability of SV40 pentamers to construct variously sized icosahedral capsids and tube-like structures makes this material extremely useful for biomedical purposes, as it means that SV40 VLPs can encapsulate a wide variety of biomaterials, including DNA, proteins, and chemicals. The pentamers can also be used to coat various artificial beads, including spherical or cuboid magnetite structures, polystyrene beads, silica beads, biodegradable poly-lactic-*co*-glycolic acid (PLGA) beads, and liposomes, regardless of their surface charge, size, and shape [35].

As mentioned above, VLPs are currently being used in biomedicine as vaccine carriers against HBV and HPV. When VLPs made from a recombinant HBV surface antigen were produced in yeast, they raised effective immune responses in humans against HBV infection. These VLPs are now commercially available as a HBV vaccine. Similarly, the VLPs of recombinant HPV L1 capsid protein that are produced in yeast or insect cells are now effective vaccines against cervical cancer. These findings suggest that the VLP platform may also be useful in other immunotherapies that aim to raise antigen-specific CTLs, helper T cells, and/or antibodies against key antigens and chemicals. For example, virus epitopes from other viruses can be inserted into the HBV or HPV VLPs: these VLPs may then induce immune responses against these epitopes that prevent infection with the foreign virus. Moreover, the VLP platform could be used to develop allergy vaccines that alter T helper responses [36]. The VLP platform could potentially also be used to treat autoimmune diseases: VLPs bearing poorly immunogenic self-antigens such as TNF-α may generate antibodies against these self-antigens that reduce the inflammatory responses in autoimmune diseases [37]. Similarly, VLPs bearing angiotensin II could reduce blood pressure while VLPs carrying amyloid-β peptide may prevent the deposition of amyloid-β and therefore provide protection from amyloidoses. Finally, the VLP platform can also be used to induce antibodies against small chemicals that prevent chemical addictions. This is demonstrated by the fact that administration of VLPs that are coupled to a nicotine derivative induces high antibody levels against nicotine and significantly increases the continuous abstinence rates of the treated smokers [38].

5.5 Technologies that modify the surface of viral capsids

Native VLPs will only infect the host cell that is targeted by the virus itself because they only have high affinity for a specific surface receptor on the host cell. However, this affinity can be changed by various technologies, thus allowing the VLP to be delivered selectively to specific cells or organs. This will improve the therapeutic efficacy of VLPs that serve as drug delivery carriers. Similarly, modifying vaccine VLPs so that they target antigen-presenting cells (APCs) may increase their immunogenicity because it will promote antigen presentation on the APCs that generate cell-mediated and/or humoral immunity. Such highly efficient vaccines will be needed when the aim is to eliminate latent or chronic infectious viruses and tumor cells from the body. Several studies show the affinities of the capsids of Ad [39], and the adeno-associated virus [40, 41] can be successfully modified by inserting a peptide ligand that has affinity for a molecule on the target cell surface, thereby generating vectors that efficiently deliver genes or therapies or serve as vaccines.

There are a number of technologies that modify the capsid surface, namely, they block the native receptor and introduce molecules that direct the VLPs to the surface of the target cell/organ (figure 5.7(C)). Some are chemical methodologies where monoclonal antibodies, proteins, and peptide/small-molecule ligands are conjugated to the VLP surface via an adapter or by cross-linking. This approach simultaneously detargets and retargets the VLPs. For example, when foreign proteins are fused to an affinity tag that interacts specifically with an external decoration protein of the VLPs of bacteriophage P22, the foreign proteins are displayed on the VLP surface [42]. Similarly, transferrin was conjugated to the surface of bacteriophage Qβ VLPs by using the azide–alkyne click reaction [43]. The click reaction was also used to immobilize a breast cancer cell-targeting peptide to the surface of hepatitis E virus VLPs [44].

It is also possible to directly insert targeting molecules in the capsid by genetic methodologies. For example, when foreign peptide sequences were inserted into the capsid proteins of an RNA phage [45] and betanodavirus [46], the peptides were displayed on the VLP surface. Similarly, when the urokinase plasminogen activator (uPA) protein was inserted genetically into the surface loops of murine polyomavirus VLPs, the chimeric VLPs selectively interacted with uPA receptor-expressing cells in a uPA receptor-dependent manner [47].

Indirect genetic approaches can also be used. For example, when the WW domain, which binds to proline-rich ligands, is inserted genetically into the surface loops of the VLP, incubation with proline-rich ligands will load these ligands onto the VLP surface [48]. A similar approach is to insert the E8C sequence (EEEEEEEEC), which interacts with the R8C sequence (RRRRRRRRC), into the surface of VLP and then incubate the VLPs with an R8C-bearing Fv fragment that recognizes the target cell surface [49]. Protein Z, an engineered antibody-binding domain, can also be inserted genetically: the resulting VLP will then be able to immobilize antibodies that recognize the target cell surface molecule. When this approach was used, the VLPs selectively bound to cells bearing the target molecule in an antibody-dependent manner [50].

We have also used various approaches to modify the surface of SV40 VLPs. First, we designed a peptide composed of three consecutive RGD motifs with three glycines on each end (GGG-RGDRGDRGD-GGG) and introduced it genetically into the DE or HI loop of SV40 VLPs [51]. The RGD motif caused the VLPs to bind to the CV-1 cell surface in an integrin $\alpha v\beta 3$-dependent manner rather than via the native SV40 receptor. We also chemically immobilized the epidermal growth factor (EGF), which is recognized by the EGF receptor (EGFR)-overexpressing cancer cells, to the VLP surface by using a hetero-cross-linker. To achieve this, we first substituted the asparagine residue at position 138 in the surface DE loop of SV40 VP1 with cysteine ($VP1^{N138C}$): this cysteine was used later to immobilize the EGF-bearing hetero-cross-linker to this position. We next fused Gaussia luciferase (GLase) or mCherry fluorescence protein (mCherry) to the C-terminus of VP2 and encapsulated it within the VLPs generated by $VP1^{N138C}$: the encapsulated fluorescence protein was used later to show whether the VLPs had entered the target cells. EGF was then attached to these VLPs by conjugating the primary amines of EGF to the succinimide group of the maleimide-succinimide hetero-cross-linker and cross-linking the resulting molecules to the new cysteine residue by using the maleimide to cysteine reaction. The resulting VLPs retained the activities of GLase or mCherry proteins encapsulated within the VLPs, which indicates that VLP encapsulation does not block the activity of the encapsulated proteins. As expected, both modified VLPs were efficiently internalized by the EGF receptor-overexpressing A431 human epidermoid carcinoma cell line, as indicated by the GLase and mCherry fluorescence. The uptake of the VLPs by the cells was EGFR-dependent.

5.6 Technologies used to encapsulate materials within the virus capsid

Various technologies (figure 5.7(D)) have been developed to encapsulate bioactive materials (proteins, chemicals, DNAs, RNAs, and polymers) into VLPs, thereby permitting the safe delivery of the materials into the target cell/organ. In terms of genetic material, various VLPs have been used as gene-delivery carriers, including those from the hepatitis E virus [52], the nodavirus [53], the papillomavirus [54, 55], the murine polyomavirus [56], SV40 [57], and human JC polyomavirus [58].

Notably, we found that when DNA is added to SV40 VP1 pentamers in physiological salt and pH conditions (150 mM NaCl, 2 mM $CaCl_2$, Tris–HCl, pH 7.0), it triggers capsid assembly and its own encapsulation [59] (figure 5.7(D), a). Thus, DNA may change the configuration of the C-terminal region of VP1 such that capsid assembly can occur. More than 750 base pairs of double-stranded DNA is needed to trigger the assembly of 45 nm-diameter capsids. When DNA with fewer than 750 base pairs is added, the resulting capsids are less than 45 nm in diameter. The maximal amount of naked double-stranded DNA that can be encapsulated is approximately 2000 base pairs. Interestingly, in the presence of more than 2000 base pairs of naked double-stranded DNA (even 116 kilo-base pairs of DNA), the pentamers still reassemble into 45 nm-diameter capsids. However, in this case, the excess DNA extrudes from the reassembled capsid. Sometimes, the same DNA molecule may bear several reassembled capsids. When we conducted a time-course study of DNA-mediated *in vitro* capsid assembly, we found that during the first 20 min after starting dialysis, the DNA only bore aggregates. However, at about the 20-minute mark, capsid reassembly suddenly started. It stopped soon thereafter, after which the aggregates had disappeared completely and reassembled capsids were instead frequently observed [59]. Thus, capsid reassembly is a very rapid reaction. These findings led us to hypothesize that nonspecific pentamer binding to DNA occurs first. Thereafter, the pentamers on the DNA rearrange themselves until an icosahedral lattice structure is formed. This will occur rapidly when the pH and salt conditions needed to trigger capsid assembly are reached, namely, 20 min after starting dialysis.

We have used this *in vitro* DNA-mediated capsid assembly reaction to generate gene-delivery carriers. First, to determine whether the VLPs generated by this system can deliver the encapsulated DNA into the cell nucleus, a plasmid encoding the enhanced green fluorescent protein (EGFP) (approximately 5000 base pairs) was encapsulated by the DNA-mediated *in vitro* capsid assembly reaction and the resulting VLPs were incubated with the African green monkey kidney cell line. Indeed, EGFP-derived fluorescence was observed within the cells [60]. By contrast, when the cells were incubated with naked EGFP-encoding plasmid, EGFP-derived fluorescence was not observed. It should be noted, however, that only very few of the cells incubated with the EGFP-bearing VLPs expressed EGFP [60]. This may have been due to the fact that the DNA protruded from the resulting VLPs because the capsid could only fully encapsulate 2000 base pairs. This may have interfered with the delivery of the DNA.

To overcome the limitation of encapsulated DNA size, we introduced the histone coordination technique. DNA is condensed by histone coordination. The SV40 virus itself uses histones to coordinate its genome, which consists of 5243 base pairs of double-stranded circular DNA: this allows this large piece of DNA to be fully encapsulated within the capsid. Thus, in a second experiment, we added a cellular histone to the naked 5000 base-pair EGFP-encoding plasmid to coordinate and condense it. This product was then used in DNA-mediated *in vitro* capsid assembly. As expected, the VLPs produced by this reaction were completely resistant to DNase I processing. By contrast, the VLPs produced by the system using the naked 5000 base-pair EGF-encoding plasmid DNA were processed by DNase I into naked DNA fragments that were approximately 2000 base pairs in size. Thus, histone arrangement compacted the DNA, thereby allowing the whole plasmid to be packed into the reassembled VLP (figure 5.7(D), b). Significantly, the reassembled VLPs containing the histone-arranged EGFP-encoding plasmid were delivered into the targeted cell nucleus much more efficiently than the VLPs that contained naked EGFP-encoding plasmid DNA [60].

This improvement was further enhanced by including VP2 and/or VP3 in the *in vitro* chromatin DNA-mediated assembly system and/or by using acetylated histone to condense the DNA. The product of chromatin DNA-mediated *in vitro* capsid assembly expressed the encapsulated DNA in the targeted cells even more efficiently than commercial transfection reagents when the same amount of plasmid DNA was used. However, the product still delivered the encapsulated DNA 100 fold less efficiently than the real SV40 virus [60]. To improve the delivery efficiency, it may be necessary to improve the histone coordination by manipulating the histone intervals and locations on the DNA: these elements significantly affect the efficiency of transcription from chromatin-coordinated plasmid DNA in the target cell nucleus [61].

A particularly intriguing potential application of VLPs is to use them to deliver small encapsulated chemicals such as drugs to target cells/organs. This will cause the drug to only affect the target cells/organs and inhibit the diffusion of the naked drug to unwanted regions of the body, thus reducing drug side effects. It is also likely that VLP-encapsulated drugs will be effective at lower doses than the naked drug, which will also help to reduce adverse drug effects. The VLPs of various viruses have been developed as drug delivery carriers. For example, Cao *et al* used the VLPs of the red clover necrotic mosaic virus (RCNMV) to encapsulate doxorubicin, a chemotherapy drug used to treat cancer. The drug was encapsulated by exploiting a unique property of RCNMV VLPs, namely, that they develop surface gaps between their VLP subunits when divalent cations are depleted or the environmental pH is increased, probably because these changes introduce conformational changes in VLP structure. These gaps allow doxorubicin to infuse into the VLPs, after which the repletion of divalent cations or reduction of pH causes the gaps to close. The VLPs can then be induced to release the drug by again depleting the divalent cations or increasing the environmental pH [62]. Similarly, Ren *et al* used hibiscus chlorotic ringspot virus VLPs to encapsulate doxorubicin. This was achieved by complexing a negatively charged polymer to the drug and incubating this complex with the coat protein monomers. Since the negatively charged polymer formed electrostatic interactions with the positively charged interior of the capsid proteins, the reassembly reaction led to efficient entrapment of doxorubicin. To ensure that the doxorubicin-encapsulating VLPs specifically targeted cancer cells, the surface of the VLPs was further conjugated with folic acid [63]. The VLPs derived from the murine polyomavirus were also used to deliver the anti-cancer drug methotrexate. To encapsulate the drug, it was covalently linked to the C-terminus of the minor capsid protein VP2 and thereby encapsulated by the *in vitro* reassembly of pentamers into VLPs. The *in vitro* internalization assay showed that the encapsulated methotrexate was delivered into human T cell leukemia cells and that the intracellularly released methotrexate was cytostatic [64]. These studies suggest that VLPs may be viable templates for targeted drug delivery in cancer chemotherapy.

Notably, Wu *et al* showed that specific chemicals can be synthesized from precursors that are encapsulated by VLPs, thus generating VLPs containing active biomaterial. For this, they first encapsulated hexacyanoferrate (III) in the VLPs of the CCMV during reassembly of the pentamers. The resulting VLPs were then reacted with a second Fe(II) to produce Prussian blue particles inside the VLPs [65]. This study suggests that VLPs can maintain their structures when chemicals such as drugs are synthesized in the VLPs.

Another intriguing potential application of VLPs is to encapsulate imaging agents within the VLPs for diagnosis: this is likely to enhance the sensitivity of the imaging agents, therefore reducing the dosage needed for imaging. This will help to prevent unwanted side effects. This possibility has been explored by several studies. For example, the VLPs of bacteriophage P22 have been used to encapsulate imaging agents. For this, the interior of the VLPs was cross-linked with a polymer that can act as a scaffold for the attachment of imaging agents such as fluorescein dye or the MRI contrast agent gadolinium-diethylenetriaminepentaacetate (figure 5.7(D), c). The resulting VLP effectively packaged these molecules [66]. Similarly, the VLPs of the human polyomavirus JC virus have been used to encapsulate propidium iodide (PI), a fluorescent dye. For this, PI was entrapped during the reassembly reaction of the VLP monomer. This procedure effectively packaged and concentrated the PI within the VLPs [67].

Many bioactive foreign proteins will not pass through the cell membrane. They also have a short retention period in the blood and do not disperse well. As a result, they largely remain at the administration site. These issues significantly limit the efficacy of these proteins. Packaging them within VLPs may help to overcome these obstacles and greatly improve their delivery since VLPs readily pass through the plasma membrane of their target cell and have high dispersibility and long retention in the blood. For these reasons, several studies have sought to develop VLPs as packaging materials for various proteins (figure 5.7(D), d). For example, the ricin A toxin was directly encapsulated in bacteriophage capsids by covalently coupling the deglycosylated ricin A chain to the assembly initiation/translational repression RNA stem-loop of the bacteriophage MS2 [68]. Similarly, the CYP450 enzyme was encapsulated in the VLPs derived from bacteriophage P22 by fusing CYP450 to a scaffold protein that aids the assembly of the VLP and interacts non-covalently with the interior of the capsid. When the CYP450-scaffold protein was coexpressed with the protein coat, 58 nm-diameter icosahedral capsids containing CYP450 were generated. When human cervix carcinoma cells were incubated with these VLPs, their CYP450 activity increased, which suggests that the VLPs had entered the cells [69]. This approach will be useful for not only delivering enzymes into target cells but also for generating nanoreactors that process specific chemicals within target cells. Several studies have shown that CCMV VLPs can also be used to encapsulate foreign proteins. This was achieved by using the *in vitro* disassembly and reassembly reaction [70]. It was also achieved by genetically adding K-coil peptides [71], or an affinity tag that is recognized by the endopeptidase sortase A (which acts to anchor the target proteins), to the capsid proteins [72]. These encapsulated proteins are likely to be useful for cell-specific diagnosis or therapies.

We found that when the minor capsid protein SV40 VP2 or VP3 (which are contained inside the native SV40 capsid) is added to the physiological buffer condition (150 mM NaCl, 2 mM $CaCl_2$, pH 7.0), the pentamers assemble into VLPs [73]. Since the pentamers alone normally do not reassemble into capsids in this condition, it seems that VP2 or VP3 functions as an acceleration factor that directs the pentamers into the VLPs. This function of VP2 is dependent on its 119–272 amino acid region (in VP3, given it is the C-terminally truncated version of VP2, the corresponding amino acid region is 1–154). Thus, this VP2 domain may change the configuration of the pentamers, thereby triggering capsid assembly. Moreover, when VP2 and/or VP3 is added to physiological salt and acidic pH conditions (150 mM NaCl, 2 mM $CaCl_2$, pH 5.0), which normally causes the pentamers to form tube-like structures, the pentamers form spherical structures. Thus, the minor capsid proteins can also correct the configuration of the C-terminal region of VP1 in these conditions. After triggering the capsid assembly, both VP2 and VP3 are encapsulated within the capsid and crystallography has shown that the C-terminal fragment of VP2 and VP3 associates with pentamers facing towards its capsid core [64, 74]. Hence, we speculated that SV40 VLPs can encapsulate foreign proteins if these proteins are fused to the C-terminus of VP2 or VP3 during VLP assembly. Indeed, Boura *et al* found that when the VP3 protein of another member of the polyomavirus family (murine polyomavirus) was fused at its C-terminus with GFP and then coexpressed, GFP was encapsulated in the VLPs and was delivered into the target cells in a VLP-dependent manner [74]. Similarly, we showed when VP2 of SV40 was fused at its C-terminus with GFP (thus generating GFP-VLPs) or yeast cytosine deaminase (yCD) (thus generating yCD-VLPs) and coexpressed it with SV40 VP1, both proteins were successfully encapsulated in the VLPs [75]. When the GFP-VLPs were administered to the cells, GFP fluorescence was seen inside these cells. By contrast, when the cells were treated with free GFP or VP2-GFP, fluorescence was only observed on the cell surface. Thus, these modified VLPs could still bind to the cell surface and enter the cell successfully. To test whether yCD activity was retained within yCD-VLP-treated cells, the ability of the cells to convert nontoxic 5-fluorocytosine (5FC) into cytotoxic 5-fluorouracyl (5FU) was examined. As expected, the cells preincubated with yCD-VLP started to die during incubation with 5FC. This suggests that the internalized yCD converted 5FC into cytotoxic 5FU in a VLP-dependent manner [75]. By contrast, the cells did not die when treated with 5FC alone or even together with naked yCD or VP2-yCD. This indicates that neither naked yCD nor VP2-yCD enters the cells and that the conversion of 5FC into 5FU in the cells does not occur. Hence, VLPs are useful for efficiently delivering encapsulated foreign proteins into cells.

5.7 Development of functional viral capsid-coated artificial beads, including magnetic nanoparticles

Yet another intriguing potential application of VLPs is to use them to encapsulate artificial beads such as gold particles, various other metallic particles, polystyrene beads, silica beads, biodegradable-polymer-based beads, and liposomes (figure 5.7(D), e). This is because these beads can be used as efficient diagnostic markers for MRI and as carriers of large amounts of drugs. Several approaches have been used to encapsulate artificial beads in VLPs. For example, metallic particles were encapsulated by the VLPs of BMV [76, 77] and CCMV [78] by simply mixing the capsid subunit with magnetite or iron oxide during *in vitro* reassembly [76–78]. Alternatively, gold nanoparticles were encapsulated in RCNMV-derived VLPs by conjugating an oligonucleotide mimic of the RCNMV RNA origin of assembly element (this element recognizes the protein coat of the virus) to the nanoparticles and then mixing the two during *in vitro* reassembly [79]. Magnetic particles were also encapsulated in HBV core VLPs by genetically introducing a histidine affinity tag into the HBV core protein, conjugating the magnetic particles with the nickel-nitrilotriacetic acid chelate, and then subjecting the two preparations to *in vitro* reassembly [80]. Moreover, gold nanoparticles were encapsulated in SV40 VLPs during *in vitro* reassembly by exploiting nonspecific electrostatic interactions between the capsid subunits and gold particles [81]. Such metallic particle-encapsulated VLPs are likely to be useful for MRI diagnostics. Furthermore, quantum dots (QDs) were encapsulated during *in vitro* reassembly of the capsid of vesicular stomatitis virus (VSV) glycoprotein pseudotyped lentivirus by conjugating them to genomic RNA that harbors the VSV packaging signal [82]. QDs were also encapsulated in SV40 VLPs by using the fact that QDs bind strongly and non-specifically to SV40 capsid subunits and thus can induce *in vitro* reassembly [83]. The resulting QD-bearing VLPs may be useful for diagnostic use. The encapsulation of metallic particles and QDs is expected to reduce bead aggregation and cytotoxicity and improve dispersal after *in vivo* administration. These properties suggest that artificial bead-bearing VLPs may also be useful as next generation drug delivery carriers for medical diagnostics and therapies.

However, how the structural properties of materials affect their encapsulation by viral capsid proteins was poorly understood until recently. To address this, we examined the encapsulation of various artificial beads by VP1 pentamers [35]. Our first experiments assessed the effect of artificial bead size. We already knew that DNA-mediated self-assembly of VP1 pentamers *in vitro* yields particles that are approximately 45 nm in diameter [59]. Moreover, we found that VP1 pentamers successfully encapsulated citrate-coated magnetic nanoparticles, which indicates that packaging can be achieved with particles whose diameter is under 45 nm [84]. Therefore, we examined whether particles that are larger than 45 nm in diameter can also be encapsulated by VP1 pentamers. Thus, VP1 pentamers were mixed with polystyrene beads with diameters of 100 or 200 nm and dialyzed the mixture against a physiological buffer (20 mM MOPS–NaOH (pH 7.0), 150 mM NaCl, 2 mM $CaCl_2$) for 16 h at room temperature. After dialysis, the mixture was washed and examined by electron microscopy. To our surprise, both the 100 and 200 nm-diameter polystyrene beads were completely covered by VP1 pentamers. To confirm that the entire surface of these beads was covered with VP1 proteins, the dialyzed sample was incubated with anti-VP1 antibody, followed by immuno-gold labeling. Indeed, the gold particles accumulated on the surface of the VP1 pentamer-coated, but not uncoated, polystyrene beads [35]. To confirm that the beads were covered with VP1 proteins, we tested the ζ potential of the polystyrene beads before and after VP1 coating. Since the ζ potential is the potential difference between the buffer and the stationary layer of fluid attached to the dispersed bead, covering the beads with VP1 pentamers will change the ζ potential of the beads to one that is closer to that of the wild-type VLP. At pH 5, the ζ potential of wild-type VLP was 10 mV, and this dropped to 30 mV at pH 9. Thus, the electron charge on the functional groups of VP1 was altered by pH. A similar change in pH did not alter the ζ potential of the uncoated polystyrene beads because the functional groups of these beads consist of phenol residues only: the electric charge of phenol residues is not altered in this pH range. By contrast, coating with VP1 pentamers altered the ζ potential of polystyrene beads, making their ζ potentials closer to those of VLPs [35].

We also assessed whether silica beads could be coated with VP1 pentamers. Unlike polystyrene beads, which are composed of phenol residues, silica beads have hydroxyl groups; the electric charge of these groups is altered by the pH. TEM showed that the silica beads were also coated with VP1 pentamers. The beads were also labeled with gold when immuno-gold labeling was performed. Similarly, the ζ potential of uncoated silica beads at pH 5–9, which was higher than that of VLP before coating, resembled the potential of VLP after coating [35]. Thus, VP1 pentamers can coat silica beads as well as polystyrene beads.

In addition, we assessed the effect of different bead shapes on VP1 coating. For this, we used cubic magnetite beads, biodegradable PLGA beads, and liposomes [35]. TEM showed that VLP treatment coated the cubic magnetite beads with a white monolayer. This was not observed with the untreated cubic magnetite beads. Since Enomoto *et al* also observed this white monolayer when citric acid-coated magnetic nanoparticles were covered with VP1 pentamers [84], it is likely that the cubic magnetite beads were also covered by VP1 pentamers. We also found that VP1 coating greatly enhanced the monodispersibility of cubic magnetite particles; by contrast, the uncoated cubic magnetite beads clustered into large aggregates. Similarly, Enomoto *et al* found that VP1-coated, but not uncoated, magnetic nanoparticles were monodispersed [84]. With regard to the PLGA beads, these uncoated beads were spherical just after preparation but formed a shriveled structure with irregular surfaces after dialysis for 16 h. By contrast, PLGA beads that were coated with VP1 pentamers maintained a spherical structure with smooth surfaces. TEM showed that liposomes, which have an irregular shape, were also successfully covered by VP1 pentamers. Thus, VP1 pentamers can coat various types of artificial beads, even those with an irregular or angular structure [35].

These findings indicate that VP1 pentamers can coat various particles without any additional steps to facilitate this attachment; this is true for beads of any size (even beads that are much larger than the wild-type SV40 capsid), shape, and ζ potential. Similarly, Li *et al* showed that QDs are efficiently coated with VP1 pentamers without needing to make specific bead modifications [85]. By contrast, when Loo *et al* used the capsid protein of RCNMV to coat gold nanoparticles, they first had to conjugate the RNA origin of the assembly element of the virus to the beads [79].

It is likely that one of the reasons that SV40 VP1 pentamers successfully coat various materials is that we used gel-filtered monomeric VP1 pentamers. Gel filtration yields two VP1 peaks: one corresponds to the void volume (>670 kDa) and one is approximately 200 kDa [29, 32, 86]. This indicates that after dissociation in the presence of EGTA and DTT, VP1 pentamers are present not only as multimeric forms but also as monomeric forms. According to the quasi equivalence theory for the development of icosahedral lattices [33], any icosahedral symmetric structure can be constructed from five-fold pentameric units and six-fold hexameric units. In particular, a $T = 7$d icosahedral lattice consists of 12 five-fold pentameric units plus 60 six-fold hexameric units. Crystallography of the SV40 capsid, which is also a $T = 7d$ icosahedral lattice, shows that the VP1 pentamers in the icosahedral lattice are present as both five-fold pentameric and six-fold hexameric units [34, 87]. The flexible ability of the monomeric VP1 pentamers that are generated by gel filtration to form both five-fold pentameric units and six-fold hexameric units on the icosahedral lattice may explain why they can easily coat variously shaped and even very large artificial beads. It is likely that the first VP1 pentamers that bind to the bead first attract other pentamers to the bead and then bind to them. As more and more pentamers bind, they adopt the five- and six-fold positions needed to accommodate each other. This process continues until the bead is completely covered with a soft net-like lattice of pentamers (figure 5.8).

In summary, VP1 pentamers can coat artificial beads of all sizes and shapes, including spherical polystyrene beads and silica beads that are larger than 45 nm in diameter, cubic magnetite beads, biodegradable PLGA beads, QDs, and even highly irregularly-shaped liposomes. The VP1 coating may also improve the monodispersibility and stability of these artificial materials. This may also be observed in body

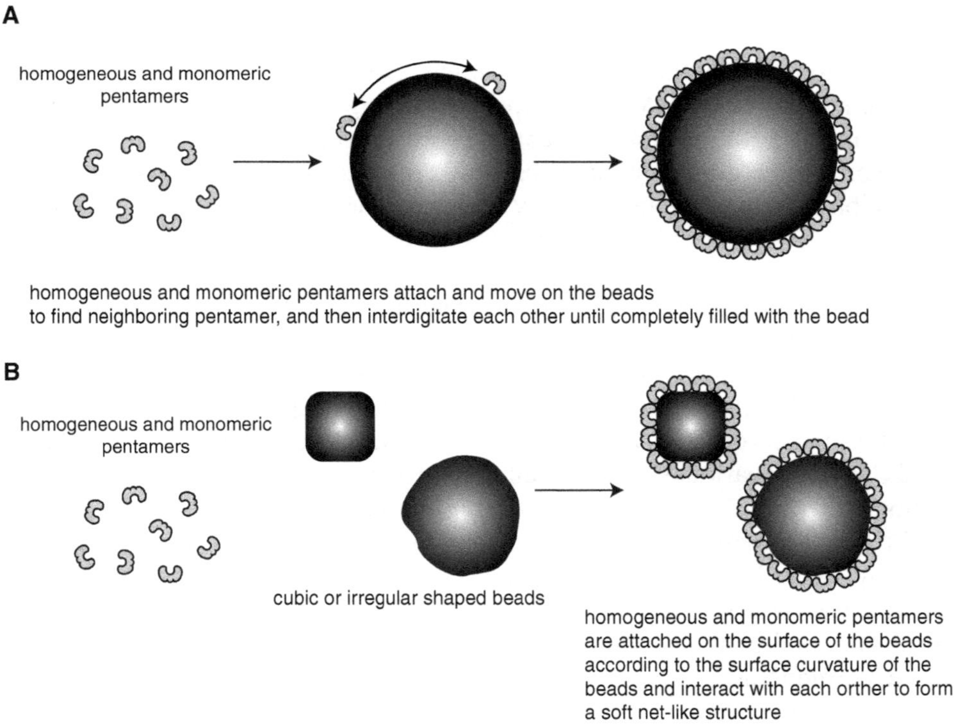

Figure 5.8. Schematic depiction of the coating of artificial beads by SV40 pentamers. (A) After a homogeneous monomeric pentamer attaches to the bead, it moves on the bead to find the neighboring pentamer. It interacts with the neighboring pentamer after finding it. This process continues until the entire surface of the bead is covered by pentamers. (B) Homogeneous monomeric pentamers can coat artificial beads of any size and shape, including angular and distorted beads, because they can have five-fold and six-fold rotational positions. This gives them the flexibility to cover all areas of the bead with a soft net-like structure.

fluids because SV40-VLPs that are composed of VP1 pentamers are highly dispersible *in vivo*, even at high concentrations. Thus, VP1 coating technology can be used with a variety of materials and may be useful for various applications in drug delivery.

5.8 Medical applications of functional viral capsid-coated artificial beads, including magnetic nanoparticles

Magnetite and fluorescence particles are, respectively, used in MRI and fluorescence imaging to visualize pathological tissues such as cancers. Other MRI contrast agents are the paramagnetic and superparamagnetic nanoparticles that are mainly composed of iron oxide or gadolinium: both types of nanoparticles reduce the relaxation time of protons [88]. All MRI contrast agents must have high dispersibility in body fluids. Moreover, since the early MRI contrast agents all diffuse well into the intravascular and interstitial spaces, there is a demand for agents that exhibit tissue-specific diffusion distribution (e.g. in malignant tumors), as this will improve the diagnostic accuracy of MRI and the ability to diagnose diseases such as cancer at an early stage [88]. Thus, the ability to target the delivery of the MRI contrast agent and induce its retention in specific tissues are also important properties.

In 2011, we reported the results of our studies, which suggested that citrate-coated magnetic nanoparticles (CMNPs) prepared by using our two-step ligand-exchange reaction method (see section 5.2) could be an excellent candidate MRI contrast agent. This was because they can be produced in various sizes (4, 8, and 20 nm) and they exhibit very good monodispersity in water. The latter property suggested that they are also likely to exhibit stable dispersibility in body fluids, which is necessary if an agent is to serve as an MRI contrast agent [21]. However, our subsequent experiments showed that contrary to our expectations, CMNPs form aggregates in physiological salt conditions.

Given that SV40-VLPs are highly dispersible *in vivo* and their VP1 pentamers readily cover all sorts of materials, we asked whether encapsulating the CMNPs with the VP1 pentamers of SV40-VLPs would improve their dispersibility. Indeed, when we mixed CMNPs of various sizes (8, 20, and 27 nm in diameter) with VP1 pentamers in a reaction buffer solution and then dialyzed the mixtures against a physiological buffer, all CMNPs of different sizes were covered with a light colored 5 nm-thick layer, as shown by TEM. By contrast, the thin layer was not observed on the surface of the untreated CMNPs. Significantly, the coated CMNPs were monodispersed in physiological salt conditions. By contrast, the untreated CMNPs formed aggregates. To confirm the monodispersion of the coated CMNPs, we measured their size distribution by dynamic light scattering. On average, the coated CMNPs were approximately 10 nm larger in diameter than the original CMNPs. Thus, CMNPs of any size can be coated with a monolayer of VP1 pentamers. By contrast, the untreated CMNPs in the solution were roughly 500 nm in diameter, which showed that they had formed aggregates. We also asked whether the VP1 coating affected the surface properties of CMNPs by measuring the electric potential of their surfaces. Due to their citrate-coat, the surfaces of the CMNPs were strongly charged at pH 7 (−18 mV). By contrast, the surfaces of the VP1 pentamer-coated CMNPs had a weak negative charge of −4 mV, which indicates that the VP1 coating changed the surface potential of CMNPs. Significantly, the surface potential of VLPs was comparable to that of the VP1 pentamer-coated CMNPs.

To assure proper assembly of SV40-VLPs, the VP1 pentamers must interact with their neighbors via their C-terminal region [25, 26]. We asked whether the C-terminal region is also important for the ability of VP1 pentamers to form a coating monolayer over CMNPs. For this, we used a VP1 mutant with a 58 amino acid deletion in the C-terminal region ($VP1^{C58}$). Thus, CMNPs were mixed with either VP1 pentamers or $VP1^{C58}$ pentamers and were then examined by TEM. The CMNPs were coated with a light colored 5 nm-thick layer regardless of whether the wild-type or mutant VP1 was used. This suggested that the pentamer–pentamer interaction is not necessary for the attachment of VP1 pentamers to the CMNP surface. However, we then found that unlike the wild-type VP1-coated CMNPs, the $VP1^{C58}$-attached CMNPs formed particulate aggregates in physiological buffer conditions. This was confirmed by sucrose gradient sedimentation analysis: of the 12 fractions that were collected from the top of the gradient after centrifugation, the wild-type VP1-coated CMNPs were detected in fractions 5–9, whereas the $VP1^{C58}$-coated CMNPs were in the bottom fraction that corresponded to the nanoparticle aggregates [73]. Thus, the C-terminal pentamer–pentamer interactions are required for the formation of dispersible VP1 pentamer-coated CMNPs.

We then tested the dispersibility of VP1-coated CMNPs in serum. Thus, VP1 pentamer-coated CMNPs and uncoated CMNPs were incubated in phosphate-buffered saline (PBS) or fetal bovine serum (FBS). The coated CMNPs were highly dispersible in both solutions whereas the uncoated CMNPs even aggregated in PBS. Dynamic light scattering analysis showed that the coated CMNPs were approximately 40 nm in diameter in both PBS and FBS; thus, the VP1 pentamer-coated CMNPs monodispersed in FBS equally well as in PBS. By contrast, the uncoated CMNPs in PBS and FBS were 400 and 150 nm in diameter, respectively. Thus, although the uncoated CMNPs dispersed readily in water, they did not disperse in either PBS or FBS.

To test whether the VP1-coated CMNPs disperse equally well *in vivo*, mice were injected intravenously with equal doses of coated or uncoated CMNPs and the concentration of the CMNP-derived Fe^{2+} in the blood was measured at various time points. Just before the injection, the baseline Fe^{2+} concentration in both cases was approximately 0.5 mg L^{-1}. Just after the injection, the mice that had been injected with the coated CMNPs had approximately two-fold higher blood Fe^{2+} concentrations than the control injected mice. The Fe^{2+} concentration of the coated CMNP-injected mice also dropped much more slowly than the Fe^{2+} concentration of the control injected mice, which indicates that the coated CMNPs were retained in the blood for a longer time. These results strongly suggest that the highly dispersible VP1 pentamer-coated CMNPs keep drifting in the blood stream, whereas the aggregated uncoated CMNPs tend to settle and are quickly taken up by the liver: the rapid uptake by the liver of these aggregates may explain why the control mice exhibited such a rapid decline in CMNP-derived Fe^{2+} in the blood.

Another increasingly important property of candidate MRI contrast agents is that they can be sent to the tissue of interest and be retained there for the duration of the MRI scan. We showed that VP1 pentamer-coated CMNPs can readily acquire this property. For this, we focused on the EGF–EGFR interaction because multiple tumor cells overexpress EGFR on their surface [89, 90]. As described above (see section 5.5), we previously constructed SV40-VLPs that bore EGF on their surface. For this, we first substituted an asparagine residue in VP1 that is located on the outer surface of VLPs [51] with a cysteine, thus generating $VP1^{N138C}$. The primary amines of EGF were then conjugated to the succinimide group of the SM(PEG)2 maleimide-succinimide hetero-cross-linker. The third step involved cross-linking the hetero-cross-linked EGF molecule to the new cysteine residue in the $VP1^{N138C}$-VLPs by using the maleimide to cysteine reaction. The resulting EGF-conjugated VLPs selectively targeted EGFR-expressing tumor cells *in vitro* [91]. Since SM(PEG)2 preferably crosslinks between a primary amine of EGF and the cysteine residue of the $VP1^{N138C}$-VLPs, EGF is probably primarily attached to the outer surface of the VLPs.

The reason we chose to use the mutated VLPs relates to the fact that while wild-type VP1 has one partially accessible cysteine residue on the surface of the VP1 pentamer, the mutation at residue 138 adds another cysteine residue to the surface loop of VP1. This may increase the surface binding of EGF to the VLPs. Mutating residue 138, which lies in the D loop of VP1 of SV40, may also block the native ability of wild-type VLPs to bind to a variety of cells via *the* ganglioside GM1 (which is recognized by the BC, DE, and HI loops of VP1) [51]. Indeed, we found that the mutation at residue 138 elicits a conformational change in the DE loop and thereby disrupts the binding of EGF-conjugated VLPs to cells that express GM1 but not EGFR. Thus, EGF-conjugated VLPs may exclusively target EGFR-bearing cells.

These data [91] and other ideas led us to purify the $VP1^{N138C}$ pentamers and use them to coat 27 nm CMNPs. The coated CMNPs were then conjugated with EGF by using the SM(PEG)2 cross-linker. The resulting $VP1^{N138C}$-CMNPs monodispersed in a physiological buffer regardless of whether they were conjugated with EGF or not. We then assessed the specificity and efficiency with which the EGF-conjugated $VP1^{N138C}$-CMNPs targeted EGFR-overexpressing cells by performing cell-binding assays with two human tumor cell lines, namely, A431, which expresses high levels of EGFR [92], and WiDr [93], which expresses low levels of EGFR. We also assessed the ability of CMNPs that were coated with wild-type VP1 and then conjugated with EGF to target these cells. To determine whether the CMNP binding to the cells was EGFR-dependent, we assessed cell binding in the presence and absence of various concentrations of free EGF. Immunoblot analysis with anti-VP1 antibody showed that both CMNP preparations bound to the A431 cells but not the WiDr cells, and that this interaction was EGFR-dependent. Notably, the A431 cells bound many more of the mutant VP1-coated CMNPs than the wild-type VP1-coated CMNPs. Thus, the mutation at residue 138 associated with more effective targeting of EGFR-expressing tumor cells. This is probably because these CMNPs bear more EGF molecules per particle. Hence, the EGF-conjugated $VP1^{N138C}$-CMNPs, and similar constructs bearing other specific tumor markers, may become useful diagnostic tools in the future.

We are currently engaged in determining whether the EGF-conjugated VP1^{N138C}-CMNPs can be used for tumor-targeting hyperthermia therapy. Since tumor cells are more heat-sensitive than normal cells, hyperthermia has been used as anti-tumor therapy [94–96]. However, given that this therapy damages normal tissues as well, targeting the heat to the tissue of interest would greatly expand the use of this therapy. Since magnetic nanoparticles can serve as heating agents in hyperthermia therapy [97], the EGF-conjugated VP1^{N138C}-CMNPs may be highly suitable for tumor-targeted hyperthermia therapy. Thus, EGF-immobilized VP1N138C-CMNPs, and similar constructs, may have great potential as a therapeutic material as well as a diagnostic material.

5.9 Conclusion

In this chapter, we described how to prepare magnetite nanoparticles and modify their surfaces for various medical uses. If these magnetite nanoparticles are to be administered to the human body for diagnostic or therapeutic purposes, they should be stable and disperse well in body fluids because rapid degradation and aggregation greatly reduce their retention period and therefore prevent them from reaching the target cell/organ. We have described two methods that greatly improve the stability and dispersibility of magnetite nanoparticles in body fluids. The first was our two-step ligand-exchange method, which greatly improves the exchange of the hydrophobic oleic acid coat on magnetite nanoparticles with hydrophilic citric acid. The second was our capsid protein coating method. When citric acid-coated magnetite nanoparticles are coated with the capsid protein, their retention in the blood is dramatically prolonged. Coating these nanoparticles with the capsid protein also promotes their efficient delivery into the target cells in the body, especially when the capsid protein is linked to a cell-targeting moiety. Magnetite nanoparticles are likely to become a useful agent for both MRI diagnosis and anti-cancer thermal therapy because of their magnetic properties and the fact that they produce heat when subjected to a magnetic field composed of specific hertz. Surface-modified nanoparticles will thus be useful for both cancer diagnosis and therapy in the future.

References

[1] Abe M, Ishihara T and Kitamoto Y 1999 magnetite film growth at 30°C on organic monomolecular layer, mimicking bacterial magnetosome synthesis *J. Appl. Phys.* **85** 5705–7

[2] Hafeli U, Schutt W, Teller J and Zborowski M 1997 *Scientific and Clinical Applications of Magnetic Carriers* (New York: Plenum)

[3] Tartaj P, Morales M P, Verdaguer S V, Carreno T G and Serma J 2003 The preparation of magnetic nanoparticles for applications in biomedicine *J. Phys. D Appl. Phys.* **36** R182–97

[4] Weaver J C, Vaughan T E and Astumian R D 2000 Biological sensing of small field differences by magnetically sensitive chemical reactions *Nature* **405** 707–9

[5] Tanimoto A, Yuasa Y, Shinmoto H, Jinzaki M, Imai Y, Okuda S and Kuribayashi S 2002 Superparamagnetic iron oxide-mediated hepatic signal intensity change in patients with and without cirrhosis: pulse sequence effects and Kupffer cell function *Radiology* **222** 661–6

[6] Nishio K *et al* 2008 Development of novel magnetic nano-carriers for high-performance affinity purification *Colloids Surf. B Biointerfaces* **64** 162–9

[7] Itoh H, Takahashi A, Adachi K, Noji H, Yasuda R, Yoshida M and Kinosita K 2004 Mechanically driven ATP synthesis by F1-ATPase *Nature* **427** 465–8

[8] Bao G and Suresh S 2003 Cell and molecular mechanics of biological materials *Nat. Mater.* **2** 715–25

[9] Le B, Shinkai M, Kitade T, Honda H, Yoshida J, Wakabayashi T and Kobayashi T 2001 Preparation of tumor-specific magnetoliposomes and their application for hyperthermia *J. Chem. Eng. Jpn.* **34** 66–72

[10] Lewin M, Carlesso N, Tung C H, Tang X W, Cory D, Scadden D T and Weissleder R 2000 Tat peptide-derivatized magnetic nanoparticles allow in vivo tracking and recovery of progenitor cells *Nat. Biotechnol.* **18** 410–4

[11] Mirkin C A and Taton T A 2000 Semiconductors meet biology *Nature* **405** 626–7

[12] Reimer P, Muller M, Marx C, Wiedermann D, Muller R, Rummeny E J, Ebert W, Shamsi K and Peters P E 1998 T1 effects of a bolus-injectable superparamagnetic iron oxide SH U 555 A: dependence on field strength and plasma concentration--preliminary clinical experience with dynamic T1-weighted MR imaging *Radiology* **209** 831–6

[13] Shinkai M, Yanase M, Honda H, Wakabayashi T, Yoshida J and Kobayashi T 1996 Intracellular hyperthermia for cancer using magnetite cationic liposomes: in vitro study *Jpn. J. Cancer Res.* **87** 1179–83

[14] Taupitz M, Schnorr J, Abramjuk C, Wagner S, Pilgrimm H, Hunigen H and Hamm B 2000 New generation of monomer-stabilized very small superparamagnetic iron oxide particles (VSOP) as contrast medium for MR angiography: preclinical results in rats and rabbits *J. Magn. Reson. Imaging* **12** 905–11

[15] Weissleder R and Mahmood U 2001 Molecular imaging *Radiology* **219** 316–33

[16] Park J, An K, Hwang Y, Park J G, Noh H J, Kim J Y, Park J H, Hwang N M and Hyeon T 2004 Ultra-large-scale syntheses of monodisperse nanocrystals *Nat. Mater.* **3** 891–5

[17] Sun S, Zeng H, Robinson D B, Raoux S, Rice P M, Wang S X and Li G 2004 Monodisperse MFe2O4 (M = Fe, Co, Mn) nanoparticles *J. Am. Chem. Soc.* **126** 273–9

[18] Jun Y W *et al* 2005 Nanoscale size effect of magnetic nanocrystals and their utilization for cancer diagnosis via magnetic resonance imaging *J. Am. Chem. Soc.* **127** 5732–3

[19] Pellegrino T, Manna L, Kudera S, Liedl T, Koktysh D, Rogach A L, Keller S, Radler J, Natile G and Parak W J 2004 Hydrophobic nanocrystals coated with an amphiphilic polymer shell: a general route to water soluble nanocrystals *Nano Lett.* **4** 703–7

[20] Wang Y, Wong J F, Teng X, Lin X Z and Yang H 2003 'Pulling' nanoparticles into water: phase transfer of oleic acid stabilized monodisperse nanoparticles into aqueous solutions of α-cyclodextrin *Nano Lett.* **3** 1555–9

[21] Hatakeyama M *et al* 2011 A two-step ligand exchange reaction generates highly water-dispersed magnetic nanoparticles for biomedical applications *J. Mater. Chem.* **21** 5959–66

[22] Iida S *et al* 2013 In vivo identification of sentinel lymph nodes using MRI and size-controlled and monodispersed magnetite nanoparticles *J. Magn. Reson. Imaging* **38** 1346–55

[23] Kawano M, Matsui M and Handa H 2013 SV40 virus-like particles as an effective delivery system and its application to a vaccine carrier *Expert Rev. Vaccines* **12** 199–210

[24] Lenz P, Day P M, Pang Y Y, Frye S A, Jensen P N, Lowy D R and Schiller J T 2001 Papillomavirus-like particles induce acute activation of dendritic cells *J. Immunol.* **166** 5346–55

[25] Garcea R L, Salunke D M and Caspar D L 1987 Site-directed mutation affecting polyomavirus capsid self-assembly *in vitro* *Nature* **329** 86–7

[26] Yokoyama N, Kawano M A, Tsukamoto H, Enomoto T, Inoue T, Takahashi R U, Nakanishi A, Imai T, Wada T and Handa H 2007 Mutational analysis of the carboxyl-terminal region of the SV40 major capsid protein VP1 *J. Biochem.* **141** 279–86

[27] Kawano M A, Xing L, Tsukamoto H, Inoue T, Handa H and Cheng R H 2009 Calcium bridge triggers capsid disassembly in the cell entry process of simian virus 40 *J. Biol. Chem.* **284** 34703–12

[28] Salunke D M, Caspar D L and Garcea R L 1986 Self-assembly of purified polyomavirus capsid protein VP1 *Cell* **46** 895–904

[29] Ishizu K I, Watanabe H, Han S I, Kanesashi S N, Hoque M, Yajima H, Kataoka K and Handa H 2001 Roles of disulfide linkage and calcium ion-mediated interactions in assembly and disassembly of virus-like particles composed of simian virus 40 VP1 capsid protein *J. Virol.* **75** 61–72

[30] Inoue T, Dosey A, Herbstman J F, Ravindran M S, Skiniotis G and Tsai B 2015 ERdj5 reductase cooperates with protein disulfide isomerase to promote simian virus 40 Eendoplasmic reticulum membrane translocation *J. Virol.* **89** 8897–908

[31] Inoue T and Tsai B 2011 A large and intact viral particle penetrates the endoplasmic reticulum membrane to reach the cytosol *PLoS Pathog.* **7** e1002037

[32] Kanesashi S N, Ishizu K, Kawano M A, Han S I, Tomita S, Watanabe H, Kataoka K and Handa H 2003 Simian virus 40 VP1 capsid protein forms polymorphic assemblies *in vitro* *J. Gen. Virol.* **84** 1899–905

[33] Caspar D L and Klug A 1962 Physical principles in the construction of regular viruses *Cold Spring Harb. Symp. Quant. Biol.* **27** 1–24

[34] Liddington R C, Yan Y, Moulai J, Sahli R, Benjamin T L and Harrison S C 1991 Structure of simian virus 40 at 3 8-A resolution *Nature* **354** 278–84

[35] Kawano M *et al* 2015 SV40 VP1 major capsid protein in its self-assembled form allows VP1 pentamers to coat various types of artificial beads *in vitro* reagardless of their sizes and shapes *Biotechnol. Rep.* **5** 105–11

[36] Moingeon P 2012 Adjuvants for allergy vaccines *Hum. Vaccin. Immunother.* **8** 1492–8

[37] Chackerian B 2010 Virus-like particle based vaccines for Alzheimer disease *Hum. Vaccin* **6** 926–30

[38] Cornuz J *et al* 2008 A vaccine against nicotine for smoking cessation: a randomized controlled trial *PLoS One* **3** e2547

[39] Hansra S, Pujhari S and Zakhartchouk A N 2015 Exploration of new sites in adenovirus Hexon for foreign peptides insertion *Open Virol. J.* **9** 1–6

[40] Liu Y, Fang Y, Zhou Y, Zandi E, Lee C L, Joo K I and Wang P 2013 Site-specific modification of adeno-associated viruses via a genetically engineered aldehyde tag *Small* **9** 421–9

[41] Michelfelder S, Varadi K, Raupp C, Hunger A, Korbelin J, Pahrmann C, Schrepfer S, Muller O J, Kleinschmidt J A and Trepel M 2011 Peptide ligands incorporated into the threefold spike capsid domain to re-direct gene transduction of AAV8 and AAV9 in vivo *PLoS One* **6** e23101

[42] Schwarz B *et al* 2015 Symmetry controlled genetic presentation of bioactive proteins on the P22 virus-like particle using an external decoration protein *ACS Nano* **9** 9134–47

[43] Banerjee D, Liu A P, Voss N R, Schmid S L and Finn M G 2010 Multivalent display and receptor-mediated endocytosis of transferrin on virus-like particles *Chembiochem* **11** 1273–9
[44] Chen C C, Xing L, Stark M, Ou T, Holla P, Xiao K, Kamita S G, Hammock B D, Lam K and Cheng R H 2016 Chemically activatable viral capsid functionalized for cancer targeting *Nanomedicine* **11** 377–90
[45] Caldeira J C and Peabody D S 2011 Thermal stability of RNA phage virus-like particles displaying foreign peptides *J. Nanobiotechnol.* **9** 22
[46] Xie J, Li K, Gao Y, Huang R, Lai Y, Shi Y, Yang S, Zhu G, Zhang Q and He J 2016 Structural analysis and insertion study reveal the ideal sites for surface displaying foreign peptides on a betanodavirus-like particle *Vet. Res.* **47** 16
[47] Shin Y C and Folk W R 2003 Formation of polyomavirus-like particles with different VP1 molecules that bind the urokinase plasminogen activator receptor *J. Virol.* **77** 11491–8
[48] Schmidt U, Rudolph R and Bohm G 2001 Binding of external ligands onto an engineered virus capsid *Protein Eng.* **14** 769–74
[49] Stubenrauch K, Gleiter S, Brinkmann U, Rudolph R and Lilie H 2001 Conjugation of an antibody Fv fragment to a virus coat protein: cell-specific targeting of recombinant polyoma-virus-like particles *Biochem. J.* **356** 867–73
[50] Gleiter S and Lilie H 2001 Coupling of antibodies via protein Z on modified polyoma virus-like particles *Protein Sci.* **10** 434–44 .
[51] Takahashi R U *et al* 2008 Presentation of functional foreign peptides on the surface of SV40 virus-like particles *J. Biotechnol.* **135** 385–92
[52] Panda S K, Kapur N, Paliwal D and Durgapal H 2015 Recombinant Hepatitis E virus like particles can function as RNA nanocarriers *J. Nanobiotechnol.* **13** 44
[53] Jariyapong P, Chotwiwatthanakun C, Somrit M, Jitrapakdee S, Xing L, Cheng H R and Weerachatyanukul W 2014 Encapsulation and delivery of plasmid DNA by virus-like nanoparticles engineered from *Macrobrachium rosenbergii* nodavirus *Virus Res.* **179** 140–6
[54] Kines R C *et al* 2015 Vaccination with human papillomavirus pseudovirus-encapsidated plasmids targeted to skin using microneedles *PLoS One* **10** e0120797
[55] Malboeuf C M, Simon D A, Lee Y E, Lankes H A, Dewhurst S, Frelinger J G and Rose R C 2007 Human papillomavirus-like particles mediate functional delivery of plasmid DNA to antigen presenting cells *in vivo Vaccine* **25** 3270–6
[56] Forstova J, Krauzewicz N, Sandig V, Elliott J, Palkova Z, Strauss M and Griffin B E 1995 Polyoma virus pseudocapsids as efficient carriers of heterologous DNA into mammalian cells *Hum. Gene Ther.* **6** 297–306
[57] Mukherjee S, Kler S, Oppenheim A and Zlotnick A 2010 Uncatalyzed assembly of spherical particles from SV40 VP1 pentamers and linear dsDNA incorporates both low and high cooperativity elements *Virology* **397** 199–204
[58] Chang C F, Wang M, Ou W C, Chen P L, Shen C H, Lin P Y, Fang C Y and Chang D 2011 Human JC virus-like particles as a gene delivery vector *Expert Opin. Biol. Ther.* **11** 1169–75
[59] Tsukamoto H *et al* 2007 Evidence that SV40 VP1-DNA interactions contribute to the assembly of 40-nm spherical viral particles *Genes Cells* **12** 1267–79
[60] Enomoto T, Kukimoto I, Kawano M A, Yamaguchi Y, Berk A J and Handa H 2011 *In vitro* reconstitution of SV40 particles that are composed of VP1/2/3 capsid proteins and nucleosomal DNA and direct efficient gene transfer *Virology* **420** 1–9
[61] Oppenheim A, Siani M, Sandalon Z and Mengeritsky G 1994 Dynamics of the nucleoprotein structure of simian virus 40 regulatory region during viral development *J. Mol. Biol.* **238** 501–13

[62] Cao J, Guenther R H, Sit T L, Opperman C H, Lommel S A and Willoughby J A 2014 Loading and release mechanism of red clover necrotic mosaic virus derived plant viral nanoparticles for drug delivery of doxorubicin *Small* **10** 5126–36
[63] Ren Y, Wong S M and Lim L Y 2007 Folic acid-conjugated protein cages of a plant virus: a novel delivery platform for doxorubicin *Bioconjug. Chem.* **18** 836–43
[64] Abbing A *et al* 2004 Efficient intracellular delivery of a protein and a low molecular weight substance via recombinant polyomavirus-like particles *J. Biol. Chem.* **279** 27410–21
[65] Wu Y, Yang H and Shin H J 2014 Encapsulation and crystallization of Prussian blue nanoparticles by cowpea chlorotic mottle virus capsids *Biotechnol. Lett.* **36** 515–21
[66] Lucon J, Qazi S, Uchida M, Bedwell G J, Lafrance B, Prevelige P E Jr and Douglas T 2012 Use of the interior cavity of the P22 capsid for site-specific initiation of atom-transfer radical polymerization with high-density cargo loading *Nat. Chem.* **4** 781–8
[67] Goldmann C, Stolte N, Nisslein T, Hunsmann G, Luke W and Petry H 2000 Packaging of small molecules into VP^1-virus-like particles of the human polyomavirus JC virus *J. Virol. Methods* **90** 85–90
[68] Wu M, Brown W L and Stockley P G 1995 Cell-specific delivery of bacteriophage-encapsidated ricin A chain *Bioconjug. Chem.* **6** 587–95
[69] Sanchez-Sanchez L, Tapia-Moreno A, Juarez-Moreno K, Patterson D P, Cadena-Nava R D, Douglas T and Vazquez-Duhalt R 2015 Design of a VLP-nanovehicle for CYP450 enzymatic activity delivery *J. Nanobiotechnol.* **13** 66
[70] Sanchez-Sanchez L, Cadena-Nava R D, Palomares L A, Ruiz-Garcia J, Koay M S, Cornelissen J J and Vazquez-Duhalt R 2014 Chemotherapy pro-drug activation by biocatalytic virus-like nanoparticles containing cytochrome P450 Enzyme *Microb. Technol.* **60** 24–31
[71] Minten I J, Hendriks L J, Nolte R J and Cornelissen J J 2009 Controlled encapsulation of multiple proteins in virus capsids *J. Am. Chem. Soc.* **131** 17771–3
[72] Schoonen L, Pille J, Borrmann A, Nolte R J and Van Hest J C 2015 Sortase A-mediated N-terminal modification of cowpea chlorotic mottle virus for highly efficient cargo loading *Bioconjug. Chem.* **26** 2429–34
[73] Kawano M A *et al* 2006 The VP2/VP3 minor capsid protein of simian virus 40 promotes the *in vitro* assembly of the major capsid protein VP1 into particles *J. Biol. Chem.* **281** 10164–73
[74] Boura E, Liebl D, Spisek R, Fric J, Marek M, Stokrova J, Holan V and Forstova J 2005 Polyomavirus EGFP-pseudocapsids: analysis of model particles for introduction of proteins and peptides into mammalian cells *FEBS Lett.* **579** 6549–58
[75] Inoue T, Kawano M A, Takahashi R U, Tsukamoto H, Enomoto T, Imai T, Kataoka K and Handa H 2008 Engineering of SV40-based nano-capsules for delivery of heterologous proteins as fusions with the minor capsid proteins VP2/3 *J. Biotechnol.* **134** 181–92
[76] Huang X *et al* 2007 Self-assembled virus-like particles with magnetic cores *Nano Lett.* **7** 2407–16
[77] Huang X *et al* 2011 Magnetic virus-like nanoparticles in N benthamiana plants: a new paradigm for environmental and agronomic biotechnological research *ACS Nano* **5** 4037–45
[78] Minten I J, Wilke K D, Hendriks L J, Van Hest J C, Nolte R J and Cornelissen J J 2011 Metal-ion-induced formation and stabilization of protein cages based on the cowpea chlorotic mottle virus *Small* **7** 911–9
[79] Loo L, Guenther R H, Basnayake V R, Lommel S A and Franzen S 2006 Controlled encapsidation of gold nanoparticles by a viral protein shell *J. Am. Chem. Soc.* **128** 4502–3

[80] Shen L, Zhou J, Wang Y, Kang N, Ke X, Bi S and Ren L 2015 Efficient encapsulation of Fe(3)O(4) nanoparticles into genetically engineered hepatitis B core virus-like particles through a specific interaction for potential bioapplications *Small* **11** 1190–6
[81] Wang T, Zhang Z, Gao D, Li F, Wei H, Liang X, Cui Z and Zhang X E 2011 Encapsulation of gold nanoparticles by simian virus 40 capsids *Nanoscale* **3** 4275–82
[82] Zhang Y, Ke X, Zheng Z, Zhang C, Zhang Z, Zhang F, Hu Q, He Z and Wang H 2013 Encapsulating quantum dots into enveloped virus in living cells for tracking virus infection *ACS Nano* **7** 3896–904
[83] Gao D, Zhang Z P, Li F, Men D, Deng J Y, Wei H P, Zhang X E and Cui Z Q 2013 Quantum dot-induced viral capsid assembling in dissociation buffer *Int. J. Nanomed.* **8** 2119–28
[84] Enomoto T *et al* 2013 Viral protein-coating of magnetic nanoparticles using simian virus 40 VP1 *J. Biotechnol.* **167** 8–15
[85] Li F, Li K, Cui Z Q, Zhang Z P, Wei H P, Gao D, Deng J Y and Zhang X E 2010 Viral coat proteins as flexible nano-building-blocks for nanoparticle encapsulation *Small* **6** 2301–8
[86] Kosukegawa A, Arisaka F, Takayama M, Yajima H, Kaidow A and Handa H 1996 Purification and characterization of virus-like particles and pentamers produced by the expression of SV40 capsid proteins in insect cells *Biochim. Biophys. Acta* **1290** 37–45
[87] Yan Y, Stehle T, Liddington R C, Zhao H and Harrison S C 1996 Structure determination of simian virus 40 and murine polyomavirus by a combination of 30-fold and 5-fold electron-density averaging *Structure* **4** 157–64
[88] Weinmann H J, Ebert W, Misselwitz B and Schmitt-Willich H 2003 Tissue-specific MR contrast agents *Eur. J. Radiol.* **46** 33–44
[89] Normanno N, Maiello M R and De Luca A 2003 Epidermal growth factor receptor tyrosine kinase inhibitors (EGFR-TKIs): simple drugs with a complex mechanism of action? *J. Cell Physiol.* **194** 13–9
[90] Salomon D S, Brandt R, Ciardiello F and Normanno N 1995 Epidermal growth factor-related peptides and their receptors in human malignancies *Crit. Rev. Oncol. Hematol.* **19** 183–232
[91] Kitai Y, Fukuda H, Enomoto T, Asakawa Y, Suzuki T, Inouye S and Handa H 2011 Cell selective targeting of a simian virus 40 virus-like particle conjugated to epidermal growth factor *J. Biotechnol.* **155** 251–6
[92] Hirabayashi K, Yano J, Inoue T, Yamaguchi T, Tanigawara K, Smyth G E, Ishiyama K, Ohgi T, Kimura K and Irimura T 1999 Inhibition of cancer cell growth by polyinosinic-polycytidylic acid/cationic liposome complex: a new biological activity *Cancer Res.* **59** 4325–33
[93] Noguchi P *et al* 1979 Characterization of the WIDR: a human colon carcinoma cell line *In Vitro* **15** 401–8
[94] Falk M H and Issels R D 2001 Hyperthermia in oncology *Int. J. Hyperthermia* **17** 1–18
[95] Koga S, Izumi A, Maeta M, Shimizu N, Osaki Y and Kanayama H 1983 The effects of total-body hyperthermia combined with anticancer drugs on immunity in advanced cancer patients *Cancer* **52** 1173–7
[96] Van Der Zee J 2002 Heating the patient: a promising approach? *Ann. Oncol.* **13** 1173–84
[97] Gupta A K and Gupta M 2005 Synthesis and surface engineering of iron oxide nanoparticles for biomedical applications *Biomaterials* **26** 3995–4021

Chapter 6

Magnetic hyperthermia treatment

Takashi Nakagawa

6.1 Introduction to magnetic hyperthermia

Recently, hyperthermia treatment is expected as cancer therapy in addition to surgery, radiation and anticancer drugs. Hyperthermia treatment in a broad sense refers to general treatment methods that cure cancer by warming the lesion part of the tumor. It has been performed as a treatment for skin tumors since the ancient Egyptian era [1]. Treatment methods of treating a target area to 47 °C or higher using infrared rays, ultrasonic waves, electromagnetic waves, etc, are called heat ablation. Treatment to warm the affected area to about 42–47 °C is called hyperthermia therapy. Keeping the temperature above 42 °C triggers the necrotic death of cancer cells. In contrast, mild hyperthermia therapy that warms to about 40–42 °C is also available. In this case, the aim is to induce the appearance of heat shock proteins that promote the improvement of immunity. Currently, applicators that utilize the energy of microwaves and RF waves are used clinically. An eddy current is generated in the body by electromagnetic waves, causing Ohmic heat generation. All body parts between the electrodes will be warmed. Because blood vessels are immature in cancer tumors, heat is easy to bask and the temperature in the tumor area rises higher than normal tissue. However, since it warms not only the cancer area but also the normal tissue, it places a heavy burden on the patient. In addition, there are problems such as an unstable effective heating technique for the lesion part of the human body and lack of reproducibility.

Magnetic hyperthermia therapy has been proposed as a method that can solve this problem. Magnetic hyperthermia therapy is a treatment method in which heating mediators that emit heat in an alternating magnetic field is introduced into the lesion part of the cancer and an alternating magnetic field is applied from the outside of the body to locally heat the cancer only. In 1957, Gilchrist *et al* first reported magnetic hyperthermia where micro size magnetite was heated in an alternating magnetic field [2]. At the beginning of 1993, Jordan and colleagues reported that magnetite nanoparticles heat well in alternating magnetic fields,

doi:10.1088/978-0-7503-1584-5ch6

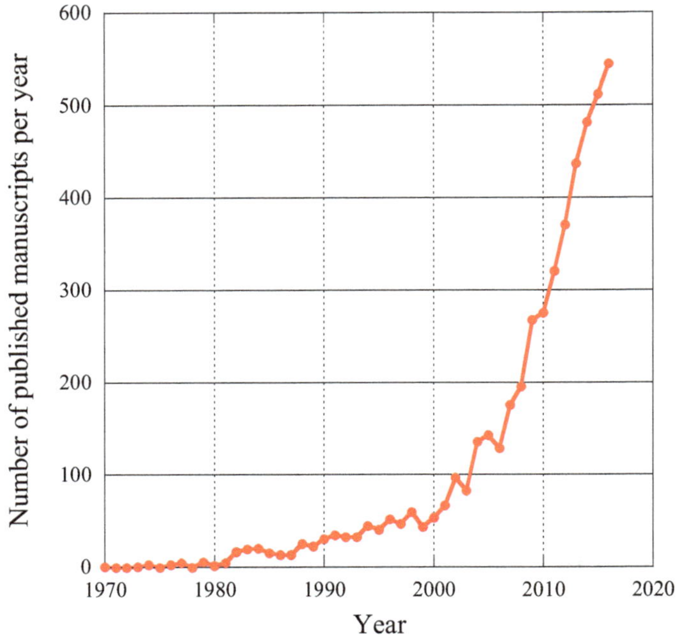

Figure 6.1. The transition of the number of paper publications each year using ‘magnetic hyperthermia’ as a search word in the literature search database Scopus.

leading to the development of magnetic hyperthermia [3]. Figure 6.1 shows the transition of the number of paper publications each year using ‘magnetic hyperthermia’ as a search word in the literature search database Scopus. From year to year, the number of papers continues to increase, showing that interest throughout the world is rising. In this chapter, the principle, current situation and future prospects of magnetic hyperthermia are explained.

6.2 Principle of magnetic hyperthermia

The frequency of the alternating magnetic field used in magnetic hyperthermia is several tens of kHz to several MHz, and it is performed under the condition that the human body does not warm due to the eddy current loss. It is known that electric conductors, ferromagnets and magnetic nanoparticles often generate heat in an alternating magnetic field of this frequency.

Electric conductors generate heat due to eddy current loss. When the electric conductor is placed in an alternating magnetic field, an induced electromotive force is generated and a current flows on the surface. The eddy current loss E at this time is

$$E = \frac{(\mu \pi d f H_0)^2}{C\rho}, \tag{6.1}$$

where μ is the magnetic permeability, d is the diameter of the particle, f is the frequency of the magnetic field, H_0 is the amplitude of the magnetic field, C is the shape dependent factor and is 20 for the sphere. ρ is the resistivity. In the case of

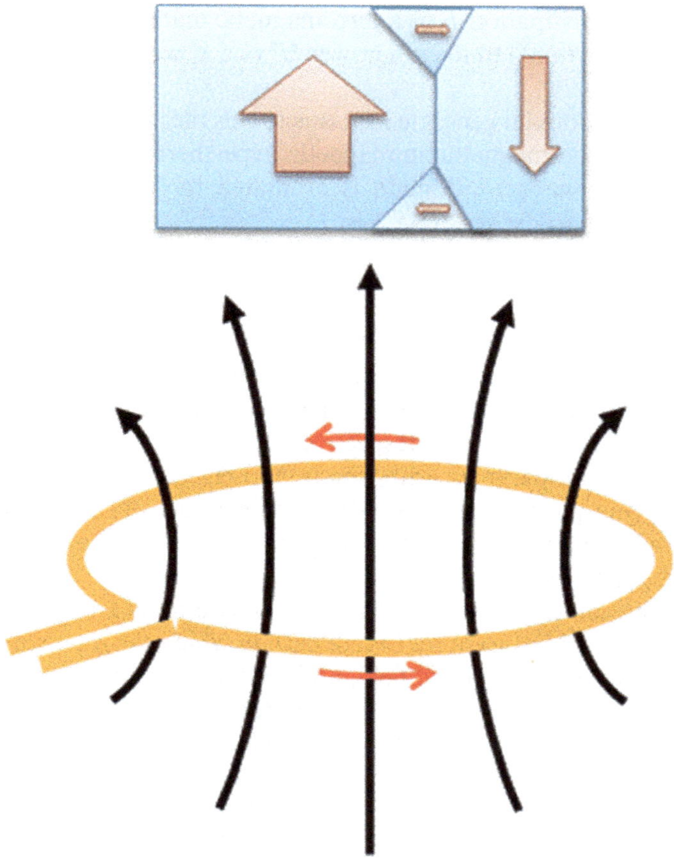

Figure 6.2. Domain wall motion of multi domain magnetic materials in an alternating magnetic field. Animation available from https://doi.org/10.1088/978-0-7503-1584-5.

magnetic nanoparticles, the eddy current loss is small due to the high resistivity, mostly neglected.

As shown in figure 6.2, the magnetization of ferromagnetic material responds to the magnetic field as the domain wall moves. Hysteresis with different magnetization occurs even with the same magnetic field strength when the magnetic field increases and decreases. The energy F stored in the magnetic material in one loop when magnetic field is applied is

$$F = \mu_0 \oint M(H)\, dH. \tag{6.2}$$

In the alternating magnetic field of the frequency f, since the loop is drawn f times per second, the heat dissipation P is expressed as

$$P = \mu_0 f \oint M(H)\, dH. \tag{6.3}$$

Generally, the heat dissipation P of a ferromagnetic material is proportional to the frequency and proportional to the nth power H^n ($1.5 \leqslant n \leqslant 2.3$) of the magnetic field strength [4].

Particles of multi domain generate heat due to this mechanism. The critical size at which the most studied magnetite in magnetic hyperthermia is changed from single domain to multiple magnetic domain is estimated to be around 130 nm [5]. In general, the critical particle size d_{critical} is

$$d_{\text{critical}} = \frac{18\sqrt{AK_1}}{\mu_0 M_s^2}, \tag{6.4}$$

where A is the exchange stiffness, K_1 denotes the anisotropic constant, M_s is the saturation magnetization [6]. For metallic nanoparticles, d_{critical} of Fe Co and Ni are reported 6 nm, 34 nm and 16 nm, respectively [7].

Ferromagnetic nanoparticles with small particle sizes are not multi-domain structures, and one particle has a single domain structure. If the single domain magnetic nanoparticles are dispersed in the matrix in the absence of interaction, they become superparamagnetic.

The magnetic moment of single domain magnetic nanoparticles rotates in two rotation mechanisms; Néel rotation (figure 6.3), in which the moment rotates in the crystal in response to the magnetic field, and Brown rotation (figure 6.4), in which the moment is fixed within the crystal and rotates as a whole. Néel relaxation time τ_N, which responds to the magnetic field by Néel rotation,

$$\tau_N = \frac{\sqrt{\pi}\,\tau_0}{2} = \frac{\exp\left(\frac{KV}{k_B T}\right)}{\sqrt{\frac{KV}{k_B T}}}, \tag{6.5}$$

where K means anisotropy constant, V is the volume of a magnetic nanoparticle, k_B is the Bolzmann constant and T means temperature. τ_0 denotes the relaxation time of non-interacting magnetic moment [8]. On the other hand, the Brown relaxation time, τ_B is expressed as

$$\tau_B = \frac{3\eta V_H}{k_B T} \tag{6.6}$$

where η is the viscosity coefficient of the matrix solution and V_H means the hydrodynamic volume of nanoparticles in the solution [9]. When τ_N and τ_B are on the same order, both rotations occur simultaneously (figure 6.5). The relaxation time τ at this time is expressed as

$$\frac{1}{\tau} = \frac{1}{\tau_N} + \frac{1}{\tau_B}. \tag{6.7}$$

The period T of the low frequency alternating magnetic field is $T \gg \tau$, and the magnetic moment can follow immediately with the change of the magnetic field.

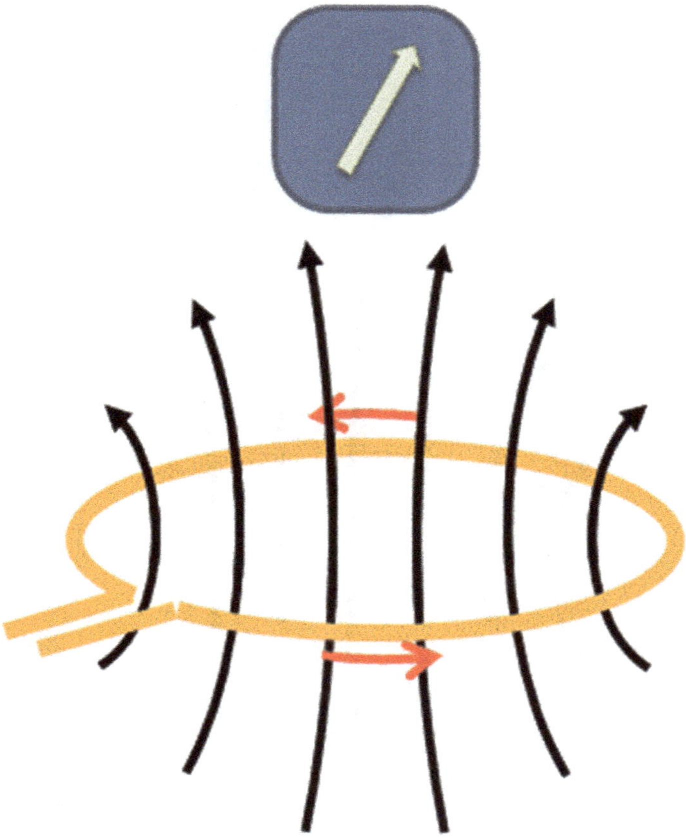

Figure 6.3. Néel rotation of a single domain nanoparticle in an alternating magnetic field with low frequency. Animation available from https://doi.org/10.1088/978-0-7503-1584-5.

Therefore, as shown in figure 6.6, the superparamagnetic magnetization magnetic field curve has the behavior of a ferromagnetic material which does not exhibit hysteresis. However, when the magnetic field period becomes shorter as compared with τ, the magnetic moment responds with a delay with respect to the magnetic field, and AC hysteresis as shown in figure 6.7 is exhibited. Figure 6.8 is a graph showing the relaxation time of the magnetite nanoparticles with respect to the particle diameter.

In 2002, Rosensweig showed the equation for the heat dissipation P of superparamagnetic magnetic fluid in an alternating magnetic field [10].

$$P = f\Delta U = 2\mu_0 f H_0^2 \chi'' \int_0^{1/f} \sin^2 2\pi f t \, dt \tag{6.8}$$

where ΔU is the cycle increase of internal energy, H_0 is maximum magnetic field strength, f means the frequency of the alternating magnetic field and χ'' denotes imaginary susceptibility. Using equilibrium susceptibility χ_0, χ'' is rewritten as

$$\chi'' = \frac{2\pi f \tau}{1 + (2\pi f \tau)^2} \chi_0. \tag{6.9}$$

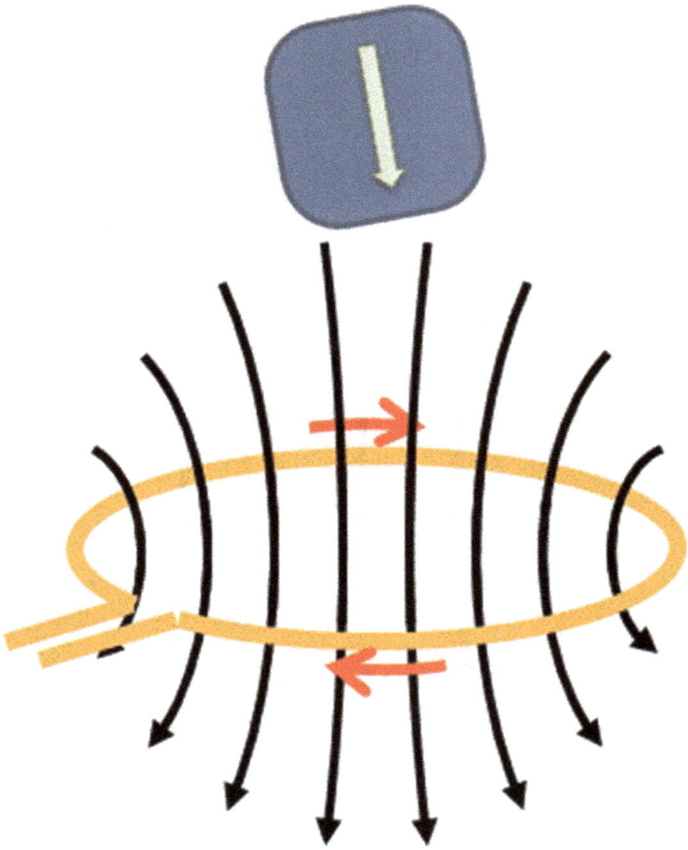

Figure 6.4. Brown rotation of a single domain nanoparticle in an alternating magnetic field with low frequency. Animation available from https://doi.org/10.1088/978-0-7503-1584-5.

When $2\pi f\tau$ is 1, the χ'' becomes maximum. Equations (6.8) and (6.9) lead to

$$P = \mu_0 \pi \chi_0 f H_0^2 \frac{2\pi f\tau}{1 + (2\pi f\tau)^2}. \tag{6.10}$$

Figure 6.9 shows the specific loss power (SLP) of monodisperse magnetite against a particle's diameter estimated from equation (6.10). As can be seen from this figure, the SLP has particle diameter dependence, and a sharp peak appears around 16 nm. In other words, it means that the optimum particle size exists in magnetic nanoparticles for magnetic hyperthermia. In actual preparation of the nanoparticles, it is difficult to synthesize them with uniform particle diameters, and a certain degree of particle size distribution occurs. The particle size distribution of nanoparticles is often represented by a lognormal distribution. The SLP, $\hat{P}$, of magnetite when the lognormal distribution is introduced is

$$\hat{P} = \int_0^\infty \mu_0 \pi \chi_0 f H_0^2 \frac{2\pi f\tau}{1 + (2\pi f\tau)^2} \cdot \frac{1}{\sqrt{2\pi}\sigma d} \exp\left[\frac{-(ln\ d - \ln d_0)^2}{2\sigma^2}\right] \mathrm{d}d. \tag{6.11}$$

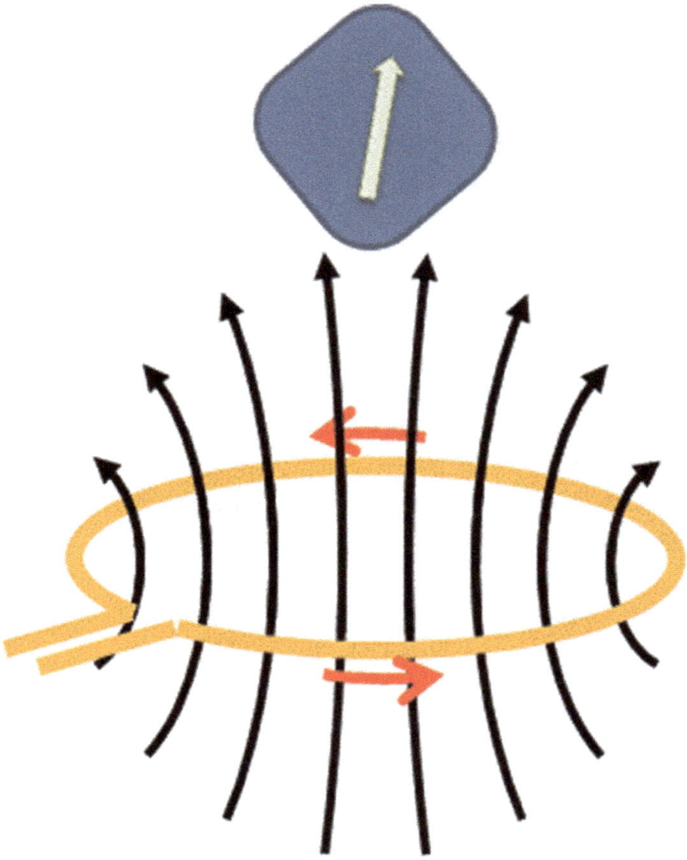

Figure 6.5. When τ_N and τ_B are on the same order, both Néel and Brown rotations occur. Animation available from https://doi.org/10.1088/978-0-7503-1584-5.

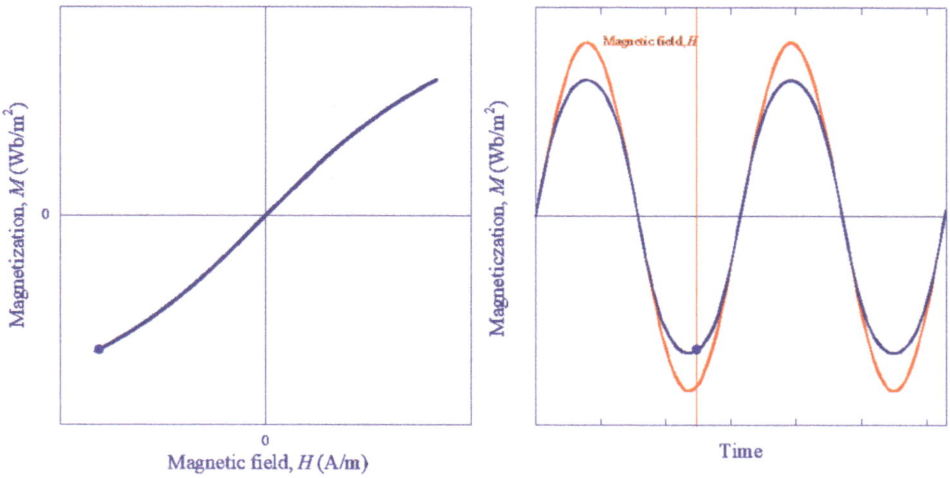

Figure 6.6. *M–H* curve of superparamagnetic nanoparticles in an alternating magnetic field with low frequency. Animation available from https://doi.org/10.1088/978-0-7503-1584-5.

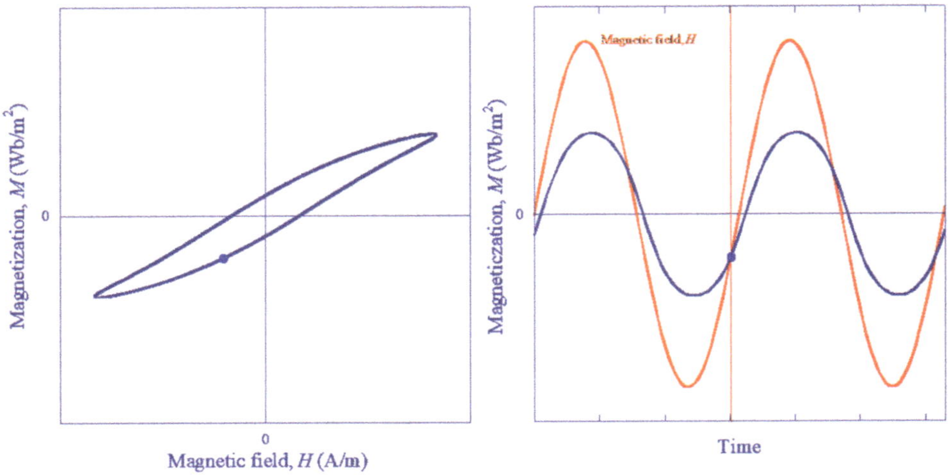

Figure 6.7. Hysteresis loop of magnetic fluid in an alternating magnetic field with high frequency. Animation available from https://doi.org/10.1088/978-0-7503-1584-5.

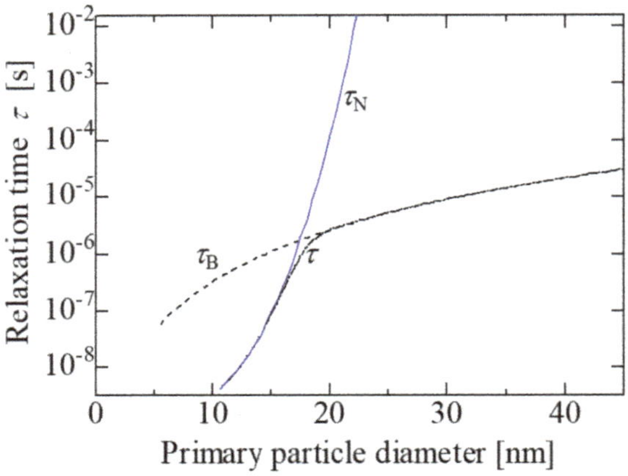

Figure 6.8. Néel relaxation time τ_N, Brown relaxation time τ_B and convolution relaxation time τ of magnetite nanoparticles as a function of particle size.

If the integral of the lognormal distribution function is normalized so that all the particles are 1 g, the heat dissipation per 1 g of particles can be calculated from the equation (6.11) when there is a particle size distribution. Since the size of the particle is often represented by the diameter rather than the radius, the diameter dependence of the SLP for various dispersions σ is also shown in figure 6.9. As you can see from the figure, the SLP decreases as σ increases. Therefore, in order to efficiently generate heat, it is necessary to use particles with a narrow particle size distribution.

6.3 Measurement of specific loss power (SLP)

SLP (or Specific Absorption Rate (SAR)) is an important index for evaluating heat dissipation characteristics of magnetic particles. SLP is a heat dissipation per unit

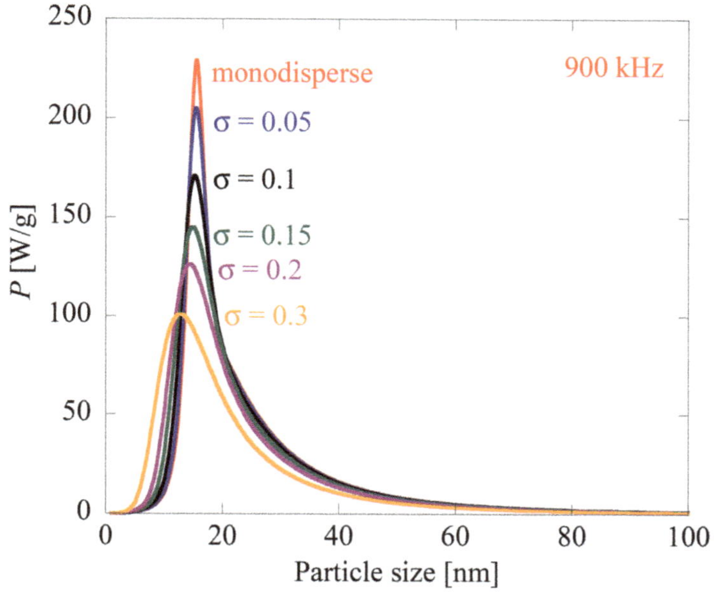

Figure 6.9. Particle size dependence of the SLP of magnetite nanoparticles with different dispersions σ.

mass and is indicated by W g^{-1} in many papers. The unit of equation (6.1) for eddy current loss is W, and equations (6.3) and (6.10) show the unit of W m^{-3}, which is divided by the density and converted into SLP. There are mainly two ways to measure SLP experimentally. One is a method of measuring temperature. The magnetic nanoparticles are covered with a thermal insulator, placed in the excitating coil, and the temperature change is observed using a fiber optic thermometer that can measure even in an alternating magnetic field. Since complete heat insulation is difficult, the temperature change saturates with temperature rise as shown in figure 6.10.

The temperature rise rate of a magnetic fluid in an AC magnetic field can be described by the phenomenological Box–Lucas equation:

$$T(t) = A(1 - e^{-Bt}) \tag{6.12}$$

where T is the temperature, t is time, A is the saturation temperature and B is a parameter related to the curvature of the heating curve [11]. The product $A \times B$ at $t = 0$ is also known as the initial heat rise rate and is equivalent to the ratio $\Delta T/\Delta t$ at $t = 0$ used in the SLP formula [12]:

$$\mathrm{SLP} = \frac{C}{m_{\mathrm{Fe}}} \frac{\Delta T}{\Delta t}\bigg|_{t=0} \tag{6.13}$$

where C is the heat capacity, and m_{Fe} is the iron mass ratio.

The second way for SLP measurement consists in measuring the dynamic magnetization $M(t)$. $M(t)$ is measured by a pick-up coil surrounding the magnetic

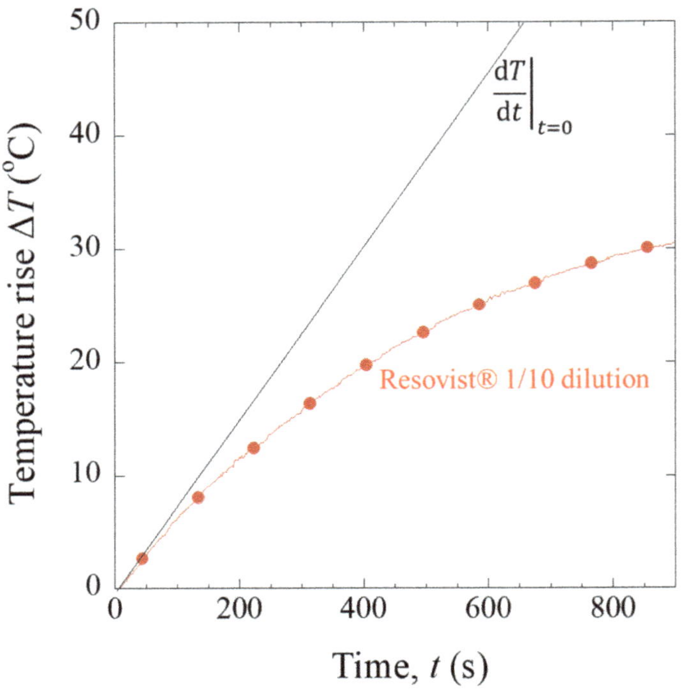

Figure 6.10. Temperature rise of 1/10 dilution Resovist® in an alternating magnetic field of 1 MHz and 4 kA m^{-1}. Temperature rise rate at $t = 0$ is also given.

fluid based on Faraday's law of induction. The voltage $v(t)$ is induced in the pick-up coil.

$$v(t) = -\frac{\partial \Phi}{\partial t} = -\frac{S \partial B(t)}{\partial t} \tag{6.14}$$

where Φ is the magnetic flux and S is the sectional area of the pick-up coil. By integrating equation (6.14),

$$B(t) = -\int \frac{v(t)}{S} dt = \mu_0 (H(t) + M(t)) \tag{6.15}$$

is obtained. In this way, $M(t)$ can be measured, and a hysteresis loop as shown in figure 6.7 can be drawn. Applying this hysteresis loop to equation (6.3), SLP is calculated.

There is intrinsic loss power (ILP) as an index representing the heat generation performance of another particle [11]. One limitation of the SLP representation is its dependence on field strength and frequency, which makes direct comparison of reported values difficult. ILP is calculated using SLP as

$$\text{ILP} = \frac{\text{SLP}}{fH^2}. \tag{6.16}$$

ILP is an indicator of the potential of heat dissipation of the material. In the case of nanoparticles with a single domain, ILP is determined according to equation (6.10). As can be seen in this equation. Magnetic nanoparticles with large χ'' have high heat dissipation potential.

6.4 Iron oxide based heating mediators

Since magnetic iron oxide nanoparticles have already been applied clinically as MRI contrast agents and have been demonstrated to have high biocompatibility, they are studied most vigorously as a heating mediator for magnetic hyperthermia [13]. Technologies covering the surface with chemicals, polymers, metals, silica, etc, have also been developed, and it is possible not only as a heating element but also has multimodal functions such as drug delivery carriers and MRI contrast agents [14–17]. Various synthesis methods for adjusting the shape and uniformity of the particle size have been reported.

The simplest method for synthesizing magnetic iron oxide nanoparticles is the coprecipitation method. By adding sodium hydroxide to iron salt solution, iron oxide nanoparticles of about 8 nm can be synthesized even at room temperature. If the reaction is carried out slowly at low temperature, the particle size increases. One of the features of the coprecipitation method is that a large amount of magnetic iron oxide nanoparticles can be synthesized. Reaction time is short. Although the particle size is relatively uniform, it is not a complete spherical shape but irregular shaped nanoparticles are formed. It is also possible to replace a part of the iron with ions such as Zn, Co, Ni and so on [18, 19].

The sol–gel method also makes it possible to synthesize magnetic nanoparticles relatively easily. This method is based on the hydrolysis and poly-condensation reactions. The sol–gel method can be used to prepare stoichiometric and mono-disperse nanoparticles. The sol–gel process is performed at room temperature, however, further heat treatments are necessary to crystalline nanoparticles [20].

Hydrothermal syntheses are based on aqueous reactions performed under high pressure using reactors or autoclaves [21]. A control of the properties of the nanoparticles is possible by choosing the appropriate mixture of solvents and reaction conditions such as temperature, pressure and reaction time. Nanoparticles with very narrow size distribution are obtained by this method.

Highly monodisperse magnetic nanoparticles are obtained by thermal decomposition of organometallic compounds such as acetylacetonates, cupferronate or carobyls in high-boiling organic solvents containing stabilizing surfactants. This process requires relatively higher temperature and complex operation [22].

As biological synthesis, it is well known that magnetotactic bacteria synthesize intercellular magnetic nanoparticles with a single domain. The magnetic nanoparticles are chemically homogeneous and enclosed by a protein embedded lipid bilayer. These nanoparticles are called magnetosome [23].

Although there have been many significant developments in the synthesis of magnetic nanoparticles, maintaining the surface properties of these particles for a long time without agglomeration is an important issue. Various materials are comprised for coating the surface of the magnetic nanoparticles. As typical polymer stabilizers, poly-ethyleneglycol, dextran, poly-vinylpyrrolidone, poly-vinyl alcohol or poly-lactic-*co*-glycolic acid are used. These polymers have been investigated for giving stealth properties to magnetic nanoparticles to enhance the blood circulation time. Oleic acid, stearic acid or lauric acid are also used as non-polymer organic stabilizers. Silica or gold plays an important role to bind various biological ligands on the surface of nanoparticles [24].

Surface modification greatly affects the heat dissipation of magnetic nanoparticles. Liu *et al* report that even the same magnetite has different SLP depending on the length of mPEG bound to their surface [25]. The shape is also one of the factors that affect SLP. The SLP of a spherical magnetite is roughly 50–200 W g^{-1} and ILP is 0.3–1.9 $nHm^2\ kg^{-1}$. Khurshid *et al* succeeded in preparing FeO/Fe_3O_4 spheres and cubes of 20 nm sizes. Although saturation magnetization of the cubes is lower than that of spheres, higher SLP of cubes is observed [26]. Lee *et al* developed magnetic nanoparticles with a magnetically hard core and soft shell, which exhibit a significant enhancement in SLP. $Zn_{0.4}Co_{0.6}Fe_2O_4$ core/$Zn_{0.4}Mn_{0.6}Fe_2O_4$ shell nanoparticles have an SLP of 3886 W g^{-1} ($f = 500$ kHz, $H_0 = 37.3$ kA m^{-1}) [27]. Noh *et al* synthesized core/shell cubes comprised of $Zn_{0.4}Fe_{2.6}O_4$ core (50 nm in edge) and $CoFe_2O_4$ shell (5 nm in thickness) showing not superparamagnetic but ferromagnetic nanoparticles. These core/shell cubes have a significantly large SLP of 10 600 W g^{-1} ($f = 500$ kHz, $H_0 = 37.4$ kA m^{-1}) [28]. This is the largest SLP reported so far in magnetic nanoparticles.

The SLP of most magnetic nanoparticles is measured in the state of solution. In measurement in solution, Brownian relaxation as well as Néel relaxation contributes to heat dissipation. Brownian relaxation loss is not relevant for the real magnetic behavior of particles because the particles do not rotate after binding to tissue. Corato *et al* have shown that SLP values shrink when magnetic nanoparticles are internalized into cells, depending on their size, shape, and coatings [29]. Thus, the SLP measurement is preferably performed in a polymer simulating an organ.

6.5 *In vivo* studies

Whether magnetic nanoparticles will rise to the temperature where treatment is effective even *in vivo* is a very important point. Numerous animal experiments have verified the effect of hyperthermia treatment of magnetic nanoparticles [30–33].

Three kinds of methods shown in figure 6.11 are considered for introducing magnetic particles into a living body; 1. Drug delivery to the cancer tumor through a blood vessel, 2. Direct injection into the tumor, and 3. Catheter is used near the tumor and the capillary of the tumor is occluded with magnetic particles of micrometer size. In the future, it will become possible to accumulate magnetic nanoparticles of sufficient concentration in the tumor part by method 1. However, at this time the amount of particles delivered to tumor tissue by antibody targeting is too low for sufficient temperature rise. As shown in figure 6.12, ideal drug transport accumulates magnetic nanoparticles only in the affected part. If this is realized, magnetic hyperthermia therapy with drug delivery becomes possible. However, as

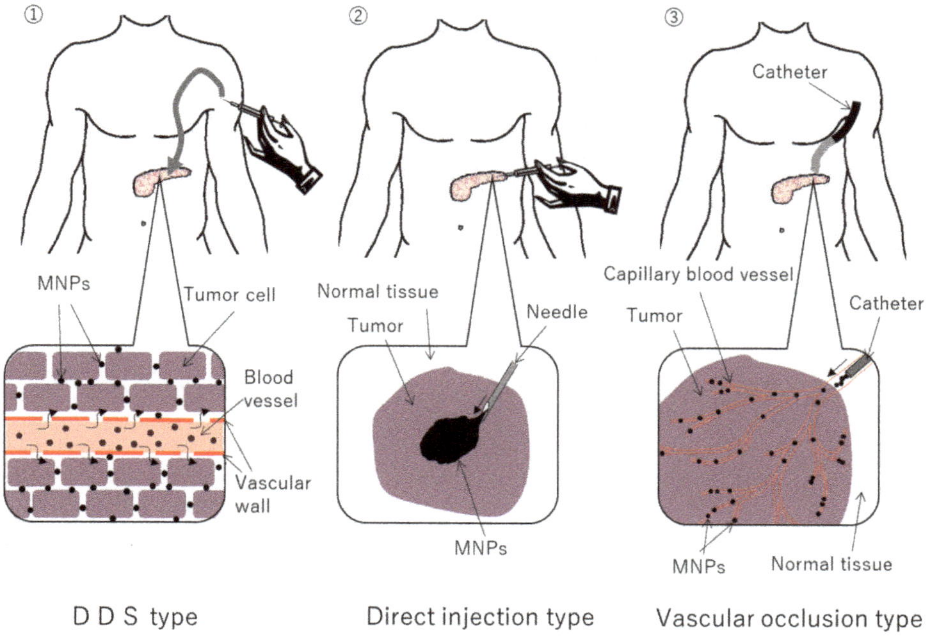

Figure 6.11. Methods for introducing magnetic particles into a living body: 1. Drug delivery system type, 2. Direct injection type, and 3. Vascular occlusion type.

shown in figure 6.13, in actual medicine transportation, although magnetic nanoparticles accumulate in the affected area, the nanoparticles of higher concentration accumulate in the liver and kidney. If an alternating magnetic field is externally applied in this situation, the liver and kidneys will be heated above the affected area. Therefore, at present, a method of directly injecting magnetic nanoparticles into a tumor is adopted.

Maier-Hauff *et al* have published the results of magnetic hyperthermia and radiotherapy with magnetic nanoparticles for 59 patients with recurrent glioblastoma. The side effects of this new therapeutic approach were moderate and no serious complications were observed. They conclude that hyperthermia using magnetic nanoparticles in combination with reduced radiation doses is safe and effective and results in longer survival compared to conventional therapy in the treatment of recurrent glioblastoma [34].

According to the website of MagForce, as of December 2017, including these 59 people,

Glioblastoma multiforme 80,
Prostate cancer 29,
Esophageal cancer 10,
Pancreatic cancer 7,
Other indications ~ 20,

clinical treatment experience with magnetic hyperthermia has been reported [35]. Future progress reports are expected.

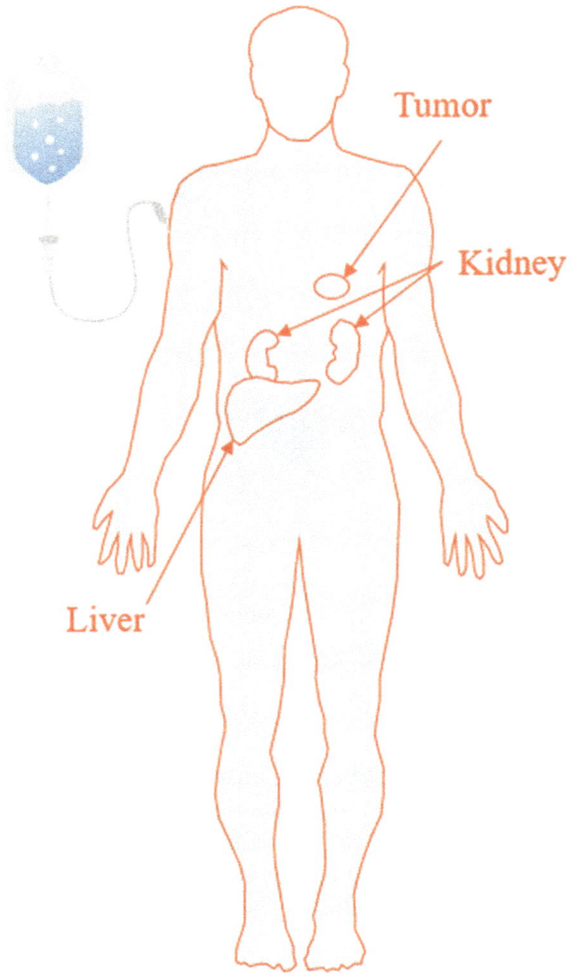

Figure 6.12. Ideal drug delivery. Magnetic particles and drugs accumulate in only the tumor area. Animation available from https://doi.org/10.1088/978-0-7503-1584-5.

6.6 Implant-type heating mediators

As compared with magnetic particles, there are very few reports on implant-type heating mediators. Implant-type heating mediators generate heat with hysteresis and/or eddy current losses. In both mechanisms, a heating mediator with a larger volume emits more heat. However, considering minimally invasive treatment, it is desirable that it can be inserted by injection, and the size of the heating mediator needs to be about 2 mm or less in diameter. For this reason, needle-like or spherical heating mediators have been developed.

Metal is a good electric conductor and generates heat due to eddy current loss in an alternating magnetic field. Since the eddy current loss is proportional to the square of the product of the frequency f of the magnetic field and the maximum

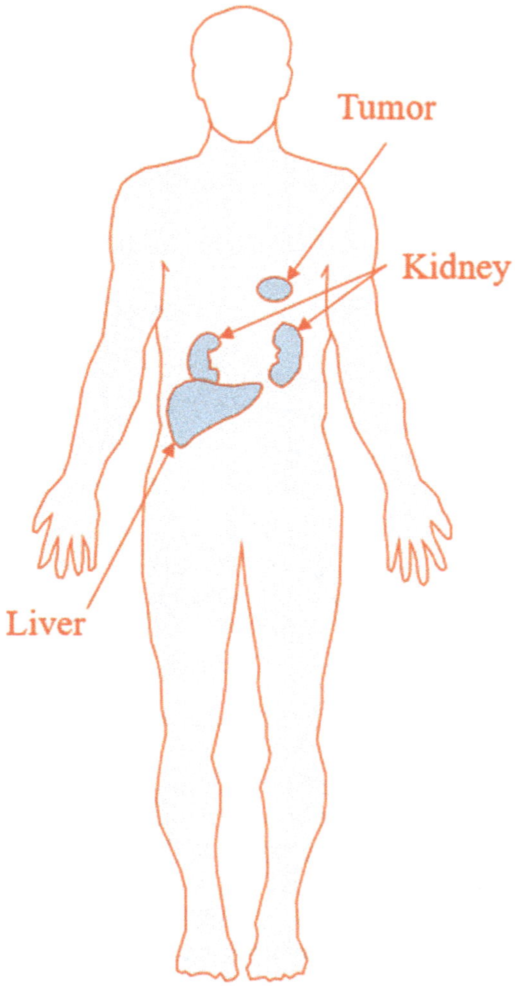

Figure 6.13. Real drug delivery. Magnetic particles and drugs accumulate in not only the tumor area but also emunctories such as liver and kidney. Animation available from https://doi.org/10.1088/978-0-7503-1584-5.

magnetic field H_0, when either the frequency or strength of the magnetic field is increased, the heat generation effect is remarkably enhanced. On the other hand, the ferromagnetic material generates heat with hysteresis loss. In the case of hysteresis loss, heat dissipation is proportional to f and is also roughly proportional to the square of H_0. For this reason, strengthening the magnetic field rather than increasing the frequency leads to an improvement in the heat generation effect.

Iron is a conductor and also a ferromagnetic material, so it generates heat with mechanisms of both eddy current and hysteresis losses. By plating gold on the surface of iron, heat generation characteristics are improved. By applying an alternating magnetic field of 300 kHz, 3.2 kA m^{-1} (rms) to 40 mg particles plated with gold having a thickness of 20 μm on an iron sphere with a diameter of 1 mm,

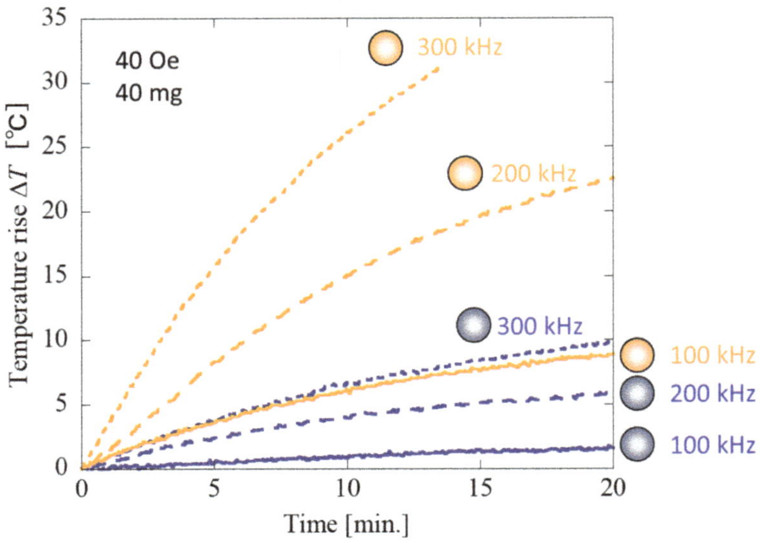

Figure 6.14. Temperature rise of starch including 40 mg of naked Fe or Au plated Fe particles with a size of 1 mm under an alternating magnetic field of 3.2 kA m^{-1} with different frequencies. The gold color means Au plated Fe and the blue color shows the naked Fe particles.

only 2 min is needed for a temperature increase of 1 mL of starch gel by 7 °C, as shown in figure 6.14.

$MgFe_2O_4$, which is a ferromagnetic material and has a large hysteresis loop, generates a large amount of heat in an alternating magnetic field. There is also a report that a heating mediator that processed $MgFe_2O_3$ into a needle shape of 1 mm in diameter was inserted into rat liver and irradiated at 540 kHz, 4.0 kA m^{-1} magnetic field for 30 min, reaching 60.7 °C [36].

However, since these heating mediators have a high Curie temperature, the temperature rises as long as the magnetic field is continuously applied, and it is difficult to accurately control the temperature of the tumor. In order to solve this problem, a method of keeping the temperature of the tumor constant by using a heating mediator having a Curie point at a temperature slightly higher than the body temperature is considered.

In 1982, Matsuki *et al* proposed a method of keeping the temperature of the diseased part constant at a temperature slightly higher than body temperature using magnetic materials, of which the Curie temperature is slightly higher than body temperature [37]. This research is the first report to control the temperature by the physical properties of the material without changing the strength of the magnetic field. Saito *et al* injected $Cu_{0.14}Zn_{0.60}Mg_{0.28}Fe_{0.98}O_3$ (Curie temperature: 43 °C) with a particle size of 100 μm into the affected part of a mouse that developed melanoma. They demonstrated that the survival rate of the mice in the treatment group is significantly higher than the control group [38]. Horiki *et al* reported that the Curie temperature can be controlled to any temperature from 41 °C to 84 °C by synthesizing $La_{2/3}Sr_{1/3}Mn_{1-x}Cu_xO_3$ perovskite, which shows a property of rapidly loss near the Curie temperature. They clarified that the temperature in the vicinity of the perovskite can be adjusted to the Curie temperature by inserting 100 mg of the bulk perovskite in 1 mL of starch gel in an alternating magnetic field [39].

Matuski *et al* reported a temperature change in a cylindrical temperature responsive magnetic material of ϕ 5 mm × 40 mm wound with a 0.1 mm thick Au ring during application of a 10 kHz AC magnetic field with an amplitude of 3–12 kA m^{-1}. As a result, it was revealed that heat generation was lost almost at the Curie temperature at any magnetic field intensity [40]. This means that if an alternating magnetic field is applied to a metal-coated heating mediator where the core is a temperature responsive magnetic material, it can be kept at a constant temperature independent of the intensity of the magnetic field. That is, even when there is a distribution in the magnetic field intensity in the body, a plurality of heating mediators can be kept at the same temperature. This is an excellent characteristic as a heating mediator for magnetic hyperthermia. By processing the temperature responsive magnetic material into a spherical shape, it is possible to obtain a heating mediator that can maintain a constant temperature without depending on the direction of the magnetic field, although the heat dissipation efficiency is inferior to that of a needle shape. Horiki *et al* prepared a spherical temperature responsive $La_{2/3}Sr_{1/3}Mn_{1-x}Cu_xO_3$ perovskite of ϕ 2.4 mm. They dispersed nine shots of perovskites (about 100 mg) in 1 mL of starch gel and measured the temperature rise of the starch gel in an alternating magnetic field of 1.6–4.4 kA m^{-1} at 1 MHz [41].

However, there are attempts to utilize the implant tool already used for cancer treatment as a heating mediator. The sealed small radiation source used for radiotherapy of prostate cancer is a ϕ 0.8 mm × 4.5 mm sized Ti capsule which seals the radiation source inside. For the treatment of prostate cancer, one can insert about 100 of this brachytherapy into about 40 mL of the prostate. It has been investigated whether a Ti needle of ϕ 0.8 mm × 4.5 mm functions as a heating mediator for magnetic hyperthermia as a cold dummy of a small radiation source. By inserting a Ti needle into a mouse that developed prostate cancer and applying a magnetic field of 990 kHz, 4.0 kA m^{-1}, it functions well as a heating mediator for magnetic hyperthermia even *in vivo* [42].

In addition, a resonant circuit type heating mediator has also been developed that generates heat with a current generated by induced electromotive force in a resonance circuit in which a wire is coiled around a magnetic core and a capacitor is built in [43]. Although it is a simulation, a method of confining magnetic nanoparticles in micron-sized capsules and using them as an implant heating mediator has also been proposed [44].

As described above, a wide variety of implant-type heating elements are also proposed at present.

6.7 Alternating magnetic field applicator

The alternative magnetic field applicator for measuring the SLP of the heating element is mostly small and the diameter of the coil is about several centimeters [45]. There is also an applicator that generates a magnetic field between the air gaps of C type ferrite cores wound with coils. The width of the air gap is several centimeters [46]. Because of their small size, these applicators are capable of generating strong magnetic fields (> 20 kA m^{-1}) at high frequencies (~ 1 MHz). Both types of applicators connect capacitors to form a resonant circuit so that an alternating current easily flows through the coil. However, a large clinical device in which a person can lie has a low driving frequency because the inductance increases. Also, since the cross-sectional area is wide, a large current is required to maintain the magnetic flux density. The German company MagForce Nanotechnologies AG has developed a large AC magnetic field generator (MFH 300FTM) that can be clinically applied. MFH 300FTM is a type of magnetic field generator equipped with a soft ferromagnetic core and it is possible to extend the air gap to 320 mm. The magnetic field frequency, f, is 100 kHz and the magnetic field strength, H, is 2.5–18 kA m^{-1} [47]. That is, the fH product of about 2 GHz A m^{-1} is a practical magnetic field condition in clinical application. Since the heat dissipation by the magnetic nanoparticles strongly depends on the strength of the magnetic field, development of magnetic field generating applicators that have uniform magnetic field intensity in the body is also desired. Takahashi *et al* have proposed an air gap type applicator capable of achieving a uniform magnetic field strength of 3800 A m^{-1} ± 5% [48].

6.8 Prospects

The properties necessary for heating mediators for clinically applying magnetic hyperthermia are summarized as follows.

1. Biocompatibility and non-toxicity.
2. Easy insertion into the body.
3. Unchanging heat dissipation characteristics with respect to the direction of the applied magnetic field.
4. Unchanging heat dissipation characteristics against magnetic field strength.
5. Sufficient heat dissipation under magnetic field condition, $fH = 2$ GHz A m^{-1}.
6. Accurate temperature control.
7. Accurate concentration control of the heating mediators accumulated in the tumor.

It is inevitable that biocompatibility and non-toxicity must be satisfied. For magnetic nanoparticles, the properties of 2 and 3 are also satisfied naturally. As for 4, it can be solved by developing an applicator capable of generating a magnetic field of uniform intensity in the space of the whole human body. A method of injecting magnetic nanoparticles directly into the tumor (by injection) has achieved the property of 5 as well. However, at the present time, the amount of particles delivered to tumor tissue by drug delivery does not provide sufficient heat. In order to realize the properties of 5 to 7, it is necessary to improve the stealth property which can be circulated for a long time in the blood and the surface modification of magnetic nanoparticles by a ligand that binds specifically to the tumor cells.

Currently, many researchers are focusing on technology development to achieve this goal, and in the near future the goal will be achieved.

References

[1] Overgaard J 1985 *History and Heritage–An Introduction In Hyperthermic Oncology* vol 2 ed J Overgaard (London: Taylor and Francis)

[2] Gilchrist R K, Medal R, Shorey W D, Hanselman R C, Parrott J C and Taylor C B 1957 *Ann. Surg.* **146** 596

[3] Jordan A, Wust P, Fähling H, John W, Hinz A and Felix R 1993 *Int. J. hyperthermia* **9** 51

[4] Tong W 2014 *Mechanical Design of Electric Motors* (Boca Raton, FL: CRC Press)

[5] Salunkhe A B, Khot V M and Pawar S H 2014 *Curr. Top. Med. Chem.* **14** 572

[6] O'Handley R C 2000 *Modern Magnetic Materials: Principles and Applications* (New York: Wiley)

[7] Skomski R 2003 *J. Phys.: Condens. Matter* **15** R841

[8] Brown W F Jr 1963 *Phys. Rev.* **130** 1677

[9] Frenkel J 1955 *The Kinetic Theory of Liquids* (New York: Dover)

[10] Rosensweig R E 2002 *J. Magn. Magn. Mater.* **252** 370

[11] Kallumadil M, Tada M, Nakagawa T, Abe M and Pankhurstab Q A 2009 *J. Magn. Magn. Mater.* **321** 1509

[12] Ma M, Wu Y, Zhou J, Sun Y K, Zhang Y and Gu N 2004 *J. Magn. Magn. Mater.* **268** 33

[13] Laurent S, Dutz S, Häfeli U O and Mahamoudi M 2011 *Adv. Colloid Interface Sci.* **166** 8

[14] Hervault A and Thanh N T K 2014 *Nanoscale* **6** 11553

[15] Lee N, Yoo D, Ling D, Cho M H, Hyeon T and Cheon J 2015 *Chem. Rev.* **115** 10637

[16] Kossatz S *et al* 2015 *Breast Cancer Res.* **17** 66

[17] Revia R A and Zhang M 2016 *Mater. Today* **19** 157

[18] Kamzin A S, Nikam D S and Pawar S H 2017 *Phys. Solid State* **59** 156

[19] Hoque S M, Tariq M, Liba S I, Salehin E, Mahmood Z H, Khan M N I, Chattopadhayay K, Islam R and Akhter S 2016 *Nanotechnology* **27** 285702

[20] Le Renard P, Lortz R, Senatore C, Rapin J P, Buchegger F, Fink A P, Hofmann H, Doelker E and Jordan O 2011 *J. Magn. Magn. Mater.* **323** 1054

[21] Chen D and Xu R 1998 *Mater. Res. Bull.* **33** 1015

[22] Lu A, Salaba E L and Schuth F 2007 *Angew. Chem. Int. Ed.* **46** 1222

[23] Hamdous Y, Chebbi I, Mandawala C, Fèvre R L, Guyot F, Seksek O and Alphandéry E 2017 *J. Nanobiotechnol.* **15** 74

[24] Salunkhe A B, Khot V M and Pawar S H 2014 *Curr. Top. Med. Chem.* **14** 572

[25] Liu X L, Fan H M, Yi J B, Yang Y, Choo E S G, Xue J M, Fan D D and Ding J 2012 *J. Mater. Chem.* **22** 8235

[26] Khurshid H, Alonso J, Nemati Z, Phan M H, Mukherjee P, Fdez-Gubieda M L, Barandiar J M and Srikanth H 2015 *J. Appl. Phys.* **117** 17A337

[27] Lee J H, Jang J T, Choi J-S, moon S H, Noh S H, Kim J W, Kim J G, Kim I S, Park K I and Cheon J W 2011 *Nat. Nanotechnol.* **6** 418

[28] Noh S H, Na W J, Jang J T, Lee J H, Lee E J, moon S H, Lim Y J, Shin J S and Cheon J W 2012 *Nano Lett.* **12** 3716

[29] Corato R D, Espinosa A, Lartigue L, Tharaud M, Chat S, Pellegrino T, Ménager C, Gazeau F and Wilhem C 2014 *Biomater* **35** 6400

[30] Ito A, Shinkai M, Honda H and Kobayashi T 2005 *J. Biosci. Bioeng.* **100** 1

[31] Hayashi K, Nakamura M, Miki H, Ozaki S, Abe M, Matsumot T, Sakamoto W, Yogo T and Ishimura K 2014 *Theranostics* **4** 834
[32] Corato R D, Béalle G, Kolosnjaj-Tabi J, Espinosa A, Clément O, Silva A K A, Ménager S and Wilhelm C 2015 *ACS Nano* **9** 2904
[33] Kossatz S, Ludwig R, Dahring H, Ettelt V, Rimkus G, Marciello M, Salas G, Patel V, Teran F J and Hilger I 2014 *Pharm. Res.* **31** 3274
[34] Maier-Hauff K, Ulrich F, Nestler D, Niehoff H, Wust P, Thiesen B, Orawa H, Budach V and Jordan A 2011 *J. Heurooncl.* **103** 317
[35] http://www.magforce.com/home.html
[36] Sato K, Watanabe Y, Horiuchi A, Yukumi S, Doi T, Yoshida M, Yamamoto Y, Tsunooka N and Kawachi K 2008 *J. Surgical Res.* **146** 110
[37] Matsuki H, Murakami L and Niizuma H 1982 *IEEE Trans. Magn.* **18** 1788
[38] Saito H, Mitobe K, Ito A, Sugawara Y, Maruyama K, Minamiya Y, Motoyama S, Yoshimura N and Ogawa J 2008 *Cancer Sci.* **99** 805
[39] Horike M, Nakagawa T, Yoshioka T, Seino S, Yamamoto T A and Abe M 2011 *J. Magn Soc. Jpn.* **35** 22
[40] Matsuki H and Murakam K 1985 *IEEE Trans. Magn.* **21** 1927
[41] Horiki M, Nakagawa T, Seino S and Yamamoto T A 2013 *J. Magn. Magn. Mater.* **329** 49
[42] Yoshioka T, Nakagawa T, Horiki M, Seino S, Yamamoto T A, Gondo T, Hashimoto T, Ohori M and Abe M 2011 *J. Magn. Soc. Jpn.* **35** 330
[43] Watabe K, Kumagai K, Matsumura R, Tamada T, Sato T and Takemura Y 2011 *IEEE Trans. Magn.* **47** 2887
[44] Zablotskii V, Lunov O and Gómez-Polo C 2010 *J. Nanosci. Nanotechnol.* **10** 1
[45] Garaio E, Collantes J M, Plazaola F, Garcia J A and Castellanos-Rubio I 2014 *Meas. Sci. Technol.* **23** 115702
[46] Seki A *et al* 2014 *J. Phys.: Conf. Series* **521** 012014
[47] Gneveckow U, Jordan A, Scholz R, Brüß V, Waldöfner N, Riche J, Feussner A, Hildebrandt B, Rau B and Wust P 2004 *Med. Phys.* **31** 1444
[48] Takahashi M, Nakagawa T, Seino S and Yamamoto T A 2014 *J. Magn. Soc. Jpn.* **38** 102

Chapter 7

Synthesis, dispersion and application of stable magnetic colloids

Balachandran Jeyadevan

7.1 Introduction

A magnetic colloid can be defined as a stable suspension dispersing magnetic nanoparticles less than 25 nm in size in either an aqueous or non-aqueous carrier fluid. A representative example is a magnetic fluid dispersing magnetite nanoparticles with an average diameter around 12 nm in organic solvents such as kerosene, toluene, etc. Particles in magnetic fluid remain stably dispersed even when they are exposed to either gravity or external magnetic field forces. Magnetic fluid constitutes a particular class of substances since they are the only fluid that can be set into motion through the sole action of the magnetic field. Thus, when a magnetic field is applied, the particles within the fluid experience a force due to the field gradient and move through the fluid imparting drag to the fluid and causing it to flow as shown in figure 7.1 [1]. When the magnetic fluid is not exposed to an external magnetic field, the suspension does not possess magnetization. The magnetic vector within each particle orients randomly such that the net spontaneous magnetization of the fluid becomes zero. However, when exposed to an external magnetic field, the magnetic moment in each nanoparticle will orient in the direction of the magnetic field lines. Though the magnetic properties of the magnetic fluid depend on the type and concentration of the magnetic nanoparticles, the flow properties strongly depend on the carrier fluid, which is the main component of the magnetic suspension. The specific magnetic properties of magnetic fluid have found very many interesting engineering applications, however, the cost and limit in the saturation magnetization of the nanoparticles have constricted their applications. On the other hand, their use in the field of fluid science as a fluid that possesses magnetism has attracted many researchers around the world, especially during the latter part of the 20th century to the beginning of the 21st century. Besides, various

doi:10.1088/978-0-7503-1584-5ch7

Figure 7.1. Behavior of iron nitride nanoparticles dispersed in magnetic fluid under an external magnetic field [1].

engineering as well as biomedical applications have been proposed and researched intensively in recent years.

7.2 History of magnetic fluid

The magnetic colloid was initially developed by Papell from NASA [2] in the early 1960s to transport fuel under zero gravity conditions. However, the magnetic particles used for the preparation of a non-aqueous based magnetic colloid was synthesized by dispersing fine magnetite particles obtained through ball milling. However, the stability of the suspension was poor even under magnetic field-free conditions and the particles settled under gravity when staying put for a prolonged period of time. Here, the colloidal suspension did not satisfy the prerequisite for 'magnetic fluid', which is defined as a suspension dispersing magnetic particles for an infinite period of time. However, particles that were small enough to be stably suspended in a fluid medium responded to the external magnetic field and exhibited novel properties. Around the same period, Satoh, Higuchi and Shimoiizaka from Tohoku University, Japan [3], developed a magnetic fluid using a completely different technique independently from Papell. Here, the magnetic particles were synthesized using a chemical technique and their diameters were around a few tens of a nanometer. Consequently, the suspension obtained was very stable over a prolonged period of time. In a true sense, the development of this stable colloid prepared by Shimoiizaka can be considered as the birth of 'magnetic fluid'.

Magnetic fluid is a complex material composed of the dispersoid (solute), carrier fluid (solvent) and dispersing agents such as ions or surfactants. Magnetic fluid can be classified as a metallic and oxide type based on the magnetic material that conforms the solute. A representative composition of the magnetic fluid is constituted of 5 vol% of magnetite as magnetic nanoparticles, 10 wt% of surfactants, and the rest is occupied by the carrier fluid [4–6]. Furthermore, the techniques to synthesize magnetite nanoparticles with average sizes around 10 nm and inhibit the aggregation among these particles are essential for the preparation of stable magnetic suspension. In the following sections, the techniques to synthesize metallic and oxide nanoparticles, the principles behind them to obtain stable suspension and the present status and future prospects on the usage of magnetic particles as well as their suspension in engineering and biomedical applications are discussed.

7.3 Synthesis of magnetic nanoparticles

Synthesis of magnetic nanoparticles has been attempted through physical vapor deposition, grinding and solution chemistry techniques. Though in the cases of both physical vapor deposition and solution chemistry techniques, the particles are formed through the assembly of atoms, the latter technique is preferred over others because molecular level control over the composition, size, etc, is easier. Consequently, solution chemistry techniques are widely used for the synthesis of metallic as well as oxide magnetic nanoparticles. As a result, control over the nucleation and growth processes has become possible through the manipulation of the synthesis parameters and thus particle size, size distribution, and shape have been controlled arbitrarily. Furthermore, solution chemistry techniques facilitate the surface modification either during or post formation of the magnetic nanoparticles and make them possess additional functions. In the following subsection, various solution chemistry techniques considered potentially viable for the synthesis of magnetic nanoparticles are introduced.

7.3.1 Synthesis of magnetic oxide particles

7.3.1.1 Co-precipitation technique

The co-precipitation technique has been used for the synthesis of metallic oxide magnetic nanoparticles for nearly half a century. In this technique, evenly mixed, solid precipitate composed of all the components is obtained by reacting metallic precursors dissolved in a solvent (generally water) with the precipitant introduced during the reaction. This precipitate undergoes dehydration to form a crystalline compound with defined sizes and shapes. This technique can be used to synthesize composite oxides of more than two metallic elements. Though the main advantage of this technique is that the process is easily scalable, control over the physical properties of the precipitate such as size and size distribution is limited due to the fact that the particle growth is hardly controllable by the reaction parameters. However, numerous spinel ferrite and perovskite type magnetic materials with wide size distributions and shapes are being synthesized using the typical aqueous co-precipitation process [7, 8]. In the case of ferrite magnetic nanoparticles, the average sizes are relatively small, around 10 nm in diameter. Besides the reaction conditions, the sizes of the nanoparticles are also influenced by the composition of the ferrites. In the context of synthesizing magnetic nanoparticles with higher magnetization, techniques to synthesize particles with diameters large enough to be dispersed in aqueous or non-aqueous solution is vital. Enhancement of the magnetic properties of ferrites has been attempted by synthesizing particles in the range of 10–100 nm utilizing the seeding technique and by introduction of various concentrations of divalent iron [9–11].

7.3.1.2 Sol–gel technique

In the conventional sol–gel process, generally, hydrolysis and the condensation of metal alkoxides are involved [12]. A metal alkoxide is a useful precursor that undergoes replacement with the hydroxyl ion from water during hydrolysis and free alcohol is formed as a by-product. Once the hydrolysis reaction is initiated, the sol is concentrated (polymerization) further by the continuation of the reaction.

The important parameters that influence the precipitate obtained through the sol–gel reaction are the solvent, temperature, precursor, catalysis, pH, additives and mechanical stirring. These parameters influence the rate of hydrolysis and condensation reactions that subsequently influences particle growth. The selection of the solvent influences the reaction rate and collision of the precursor. And the pH influences the hydrolysis and condensation reactions. Acidic conditions facilitate the hydrolysis reaction. It should be noted that the hydrolysis products are formed prior to the condensation process. Also, cross-linking density is low under acidic conditions and the density of the final product becomes high when the gel is broken. Under alkaline conditions, the condensation reaction progresses and in contrast to the acidic conditions the condensation reaction occurs prior to the completion of the hydrolysis reaction. Furthermore, the pH of the solution influences the isoelectric point and the stability of the sol, in other words, the aggregation and the final size of the particle is influenced a lot. Thus, by varying the conditions that influence the reaction rate of hydrolysis and condensation, the structure and properties of the gel can be controlled. Although these reactions progress at room temperature, heat treatment is necessary to obtain crystalline particles. However, since the particles obtained during synthesis are either in the amorphous or metastable state, sintering of the particles through annealing can be achieved at temperatures lower than required for normal solid phase.

For the syntheses of multicomponent particles with either two or three elements, the sol–gel method is more attractive than the co-precipitation technique when the preparation of metal hydroxide from a metallic precursor solution is considered. Here, by considering either a molecule composed of several metallic elements or composite alkoxide, the synthesis of multicomponent particles becomes possible. The syntheses of numerous ferrite nanoparticles has been reported using this technique [13, 14].

7.3.1.3 Micelle technique

Surfactants spontaneously form a micelle or a spherical aggregate called 'micro-emulsion' in a solvent. In a micelle, the hydrophilic end of the surfactant faces outside the micelle aggregate and in reverse the micelle hydrophobic end faces outside. In other words, the micelle forms irrespective of the existence of water. In the case of a micelle formed in a hydrocarbon based solvent, water is dissolved in hydrocarbon forming a 'water pool' and the size of this is determined by the ratio between the water and surfactant. The particle growth using the above technique is kinetically and thermodynamically controlled by the 'nanoreactor'. In reports on the synthesis of metallic and oxide nanoparticles, aerosol-OT, cetyltrimethylammonium bromide (CTAB), sodium dodecyl sulfate, polyethoxylate are used as surfactants [15–17]. The crystal structure of the metallic particles is reported to be influenced by the surfactant types [15–18]. A detailed account on the synthesis of metallic nanoparticles is also reported [19].

7.3.1.4 Thermal decomposition technique

Compared to the conventional co-precipitation technique to fabricate magnetic oxides in the 1970s [7], a modern synthetic approach, such as thermal decomposition of metal precursors in high boiling point organic solvents, has been developed. This technique has been well demonstrated in the case of iron oxide nanoparticles and in most cases, nanoparticles with controllable size and shape have been obtained [20–25, 27–32]. Though this technique was developed by Hyeon and his co-workers by thermally decomposing iron oleate in the presence of oleic acid to obtain particles ranging between few nm to few tens of nm [33 26], this technique has been extended to other metallic oxides. Furthermore, non-spherical cubic, concave, and multi-branched shaped particles using dual-metal oleate complexes as a precursor, have been achieved [27, 28, 34, 35]. Figure 7.2 demonstrates the ability of the thermal decomposition technique to produce size- and shape-controlled synthesis of magnetite particles [29]. In addition, the variation in the surfactant type used during the synthesis has added another dimension to the type of particles prepared using this technique. Though the thermal decomposition technique is not suitable for large-scale synthesis compared to the conventional co-precipitation technique, it has become a prominent method among researchers working on the physics of magnetic nanoparticles.

7.3.2 Synthesis of magnetic metal particles

7.3.2.1 Thermal decomposition technique

The syntheses of metallic nanoparticles through the decomposition of organo-metallic precursors by heat (thermolysis) [30, 31], light (photolysis) [32], or sound (sonolysis) [33] is often performed. Since the properties of the nanoparticles are influenced by the size and shape, it is necessary to control the particle growth by tuning the reaction temperature and surface modifying agents, besides other parameters that influence the overall reaction. In most cases, polymer or organic surface modifying agents are used to control the nanoparticle growth. There are

Figure 7.2. Magnetite particles with different sizes and shapes using the thermal decomposition technique [29].

stabilizing polymers that influence the formation of spherical shaped particles usually obtained by the thermal decomposition of metal carbonyls [34]. The thermal decomposition technique is being extensively used for the synthesis of metallic and alloy nanoparticles. However, the sizes of the particles obtained though this technique have been limited due to the high reactivity of the system leading to the formation of a large number of nuclei that limits subsequent growth [35].

7.3.2.2 Borohydride reduction method

The synthesis of monodispersed spherical shaped metal iron, cobalt and nickel nanoparticles using sodium borohydrate as a reductant has been reported [36–38]. Though the formation of boride as a by-product during synthesis has been considered an issue, the control of the reaction atmosphere and the water content is reported to inhibit its formation. Pure phase materials can be synthesized by using non-aqueous solvents such as tetrahydrofuran, dimethylglyoxime, etc. The additional merit of this process is that by mixing a number of metal salt types prior to the reduction reaction could lead to the formation of alloy nanoparticles relatively easily [39].

7.3.2.3 Polyol process

The polyol process is a synthesis technique to obtain metallic powders by the reduction of inorganic compounds in liquid polyols such as ethylene glycol. Here, polyol acts as a solvent, reducing agent, and surfactant, and the metallic and alloy particles with controlled sizes and crystal structures can be synthesized by manipulating the reaction parameters and additives [40–42]. However, most of the metallic and alloy magnetic particles are in the range of sub-micron meters and the synthesis of nanometer-sized particles has been a challenge. In recent years, the successful synthesis of metallic nickel nanoparticles has been reported, but the synthesis of nanosized iron and cobalt particles continues to be challenging. As a non-aqueous solvent, polyol has a high dielectric constant due to the possession of hydrogen bonds and consequently has the ability to dissolve inorganic compounds. Metallic particles can be synthesized by using the polyol process by dissolving metallic sulfates, chlorides, acetates or oxides in diols such as ethylene glycol, trimethylene glycol or tetraethylene glycol and heating the solutions at the boiling point of the respective solvents under refluxing conditions. The dissolution of the metal precursors becomes possible in diols during the reaction and intermediate compounds are formed. As the reaction progresses, the intermediate subsequently is reduced to form metallic particles. In addition to easily reducible noble metals such as Ru, Rh, Pd, Ag, Os, Ir, Pt and Au, the synthesis of transition metals such as Co, Ni, Cu and Pb has also been reported [43]. Furthermore, the application of this technique has been extended to metal composite powders composed of metals that are not easily reduced, such as iron. The size of the nanoparticles depends mainly on the synthesis conditions such as metal ion concentration, polyol type and reaction temperature, besides other parameters. Further tuning of the particle size can be achieved though the introduction of seed particles [40–45].

At the initial stages, most of the magnetic particles obtained using this technique were in the sub-micron range. However, in the recent past, additives and nucleating agents have made control over the reaction rate possible and the synthesis of nickel, cobalt and iron nanoparticles has been realized. Consequently, the application of such materials in the field of engineering has been on the rise [46–54].

In the context of the synthesis of metal and alloy magnetic nanoparticles, the polyol process possesses many excellent features compared to other non-aqueous techniques such as thermal decomposition, etc, and can be considered as a practical solution to magnetic nanomaterial synthesis [45, 49, 52]. However, in order to exploit the full potential of this process, better understanding on the reduction/oxidation reaction schemes accounting for the formation of metallic nanoparticles should be elucidated. Recently, considerable progress has been made in elucidating the active species responsible for the reduction of metal ions [55] and also the redox reaction in ethylene glycol–cobalt system [56]. This has made the visualization of the reactions that are taking place during the synthesis of metals and alloy nanoparticles possible. Also, research is in progress to formulate the criteria for the selection of metal salts using the knowledge of general chemistry, such as solubility products, etc, and the type of polyols using calculation chemistry. Consequently, the synthesis of metallic cobalt and iron particles with sizes less than 100 nm has been achieved [57] (figure 7.3). It is strongly believed that the above finding will make the synthesis of magnetic metallic iron particles in the nanosize range possible in the near future.

In this section, the outline of various synthesis techniques applicable for the synthesis of magnetic nanoparticles have been introduced. Besides magnetic nanoparticles, research findings related to various inorganic particles and magnetic

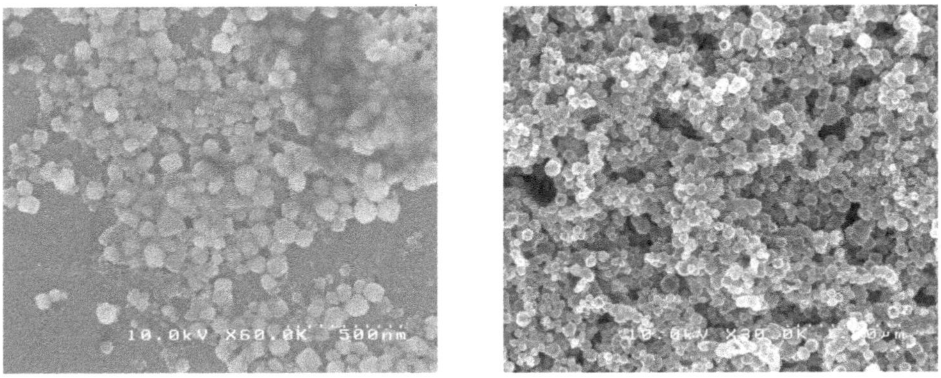

(a) (b)

Figure 7.3. Scanning electron micrographs of metallic (a) Co and (b) Fe nanoparticles synthesized by the polyol process [55].

particles whose sizes range between a few hundred to a few microns are also available. Those who are interested in the above area are requested to refer to recent reports [58–60].

Here, a few potential techniques for the synthesis of magnetic nanoparticles have been introduced. In the case of metallic oxides, the co-precipitation technique is widely used. Though this technique is suitable for large-scale synthesis, control over properties such as particle size, size distribution, etc, is considered difficult. On the other hand, the solution for size and size distribution control could be found in the thermal decomposition techniques proposed in recent years. However, the inability of this technique to synthesize metallic oxides on a large-scale remains a challenge.

However, thermal decomposition is also being used to synthesize metallic nanoparticles with reasonably narrow size distribution, by controlling the reaction atmosphere. However, the evolution of carbon monoxide during the thermal decomposition of metal carbonyl and the upper size limit of the particles that could be obtained using this technique limits their applicability.

In contrast to the above technique, the synthesis of non-magnetic and magnetic metallic and alloy particles [61] (figure 7.4) using the polyol process has been on the rise and is very promising. Recent progress in the understanding of the redox reaction and the particle formation mechanism has enhanced the potential of this process more than ever and the development of highly functional magnetic nanoparticles can be greatly anticipated.

7.4 Stability of magnetic fluid

7.4.1 Theory

To maintain the stability of the magnetic fluid, first the surface of the nanoparticle should be modified to prevent the aggregation due to the attractive energy between them caused by the van der Waals force.

Figure 7.4. SEM micrograph of FeCo nanoparticles synthesized using the polyol process [61].

In addition, the size of the nanoparticle should be small enough such that the Brownian motion energy of the particle becomes greater than the attractive magnetic energy between particles dispersed in the solvent and gravitational energy. Consequently, the stability of the magnetic fluid is determined by the balance between thermal energy, attractive (van der Waals and dipole–dipole interaction) and repulsive forces. The critical diameter that inhibits the magnetic aggregation of particles can be estimated by comparing the thermal energy and dipole–dipole interaction as follows [62].

$$D < (72k_BT/\pi\mu_0M^2)^{1/3} \tag{7.1}$$

Here, k_B is Boltzmann constant, T, the absolute temperature, μ_0, permeability in vacuum. M is the intensity of magnetization. The critical diameter for a particle dispersed in either water or organic solvent estimated using the above equation becomes smaller than 10 nm.

The magnetic particle dispersed in magnetic fluid mainly receives van der Waals and dipole–dipole interaction forces. These forces are effective at short distances and increase as the particle diameter increases. The van der Waals attraction, U_{Av} for particles with diameter and D separated by a distance of r, is given by the following equation.

$$U_{Av} = -A/6\{[2/(a^2 - 4)] + (2/a^2) + \ln[(a^2 - 4)/a^2]\};\ a = 2r/D \tag{7.2}$$

Here, A is the Hamaker constant (~10^{-19} J in the case of ferrites) [63]. As understood from this equation, the van der Waals interaction, U_{Av} is inversely proportional to the sixth power of the inter-particle distance and becomes negligible when the particles are a distance apart. On the other hand, it increases suddenly when the particles approach each other and lead to agglomeration that destabilizes the dispersion of the particles.

However, the magnetic interaction U_{Am} between particles with permanent dipole magnetic moments $\boldsymbol{\mu}_1$ and $\boldsymbol{\mu}_2$ at a relative distance of $\mathbf{r}$ is given by the following equation.

$$U_{Am} = 1/\mu_0 4\pi r^3\{\boldsymbol{\mu}_1 .\ \boldsymbol{\mu}_2 - 3[\mu_1 .\ (\mathbf{r}/r)][\mu_2 .\ (\mathbf{r}/r)]\} \tag{7.3}$$

Considering the diameter of the magnetic particle to be around 10 nm, separated by a distance of about 2 nm, then the magnitude of the magnetic interaction becomes negligible at room temperature. However, when the diameter of the particles becomes larger than 15 nm, the magnetic dipole interaction becomes strong enough to induce agglomeration between particles that could lead to instability of the dispersion. Consequently, due consideration should be given to keep the particles farther apart compared to particles less than 15 nm.

Thus, even in the case of magnetic fluid, the generation of repulsive force between particles is generated either through the steric repulsion between surfactants adsorbed on their surface or by the electric repulsion due to surface electric double layer. The steric repulsive force has a linear relationship with temperature. The repulsive energy, U_{Rs} for a surfactant thickness, δ, molecular density per square nm, ζ, and temperature, T is given by the following equation:

$$U_{Rs} = (\pi D^2 \zeta k_B T)/2\{2 - [(l + 2)/t]\ln[(l + t)/(1 + l/2) - l/t]\} \tag{7.4}$$

where $l = 2\,(r - D)/D$, $t = 2\delta/D$ (D, diameter of the particle). On the other hand, in the case of the adsorption of highly polarized ions, the driving force for the repulsion between particles is due to the long distance electrostatic interaction.

However, when the dispersion is achieved through high polar ions, the driving force for repulsion is the long distance electrostatic interaction. In the case of ionic magnetic fluid developed by Massart *et al* [64] control over the concentration of the surface potential determining ions such as H^+ or OH^- was realized by selecting the appropriate acid or alkali coordinating ions. The surface potential of magnetite becomes zero (pzc, point of zero charge) at some specific pH value. And at pH values higher than pzc, the surface of magnetite becomes negatively charged and becomes positively charged at lesser values than pzc. To neutralize the surface charge, counter ions accumulate and distribute in the vicinity of the interface. If the chemical interaction between these counter ions and the hydration layer is very small, then the behavior is determined solely by the electric interaction and as a result the interface potential becomes almost equal to the surface potential.

In an aqueous system, the particle becomes charged due to deficiency in the surface charge caused by the surface dissociation of the components. Since this is included in the surface double layer, the electric repulsion force acts in the vicinity of the particle. This electric repulsion energy is given by [65]

$$V_R = [D\pi\sigma^2/\varepsilon_0\varepsilon_\rho\kappa^2]\exp[-\kappa(r - D)] \tag{7.5}$$

Here, the surface charge density σ is $\varepsilon_0\varepsilon_p\kappa\psi_0$, the dielectric constant of the solvent ε is $\varepsilon_0\varepsilon_p$ and κ is the thickness of the ion atmosphere. Also, ψ_0 is the surface charge of the charged particle at the Helmholtz surface. If equation (7.5) is used instead of equation (7.4), the state of dispersion on systems that undergoes stabilization through the repulsion force generated by the electric double layer can be analyzed. Furthermore, even in the case where the magnetic interaction between particles is assumed to be zero, the surface potential has to be more than 50 mV to maintain the peak height of the energy potential curve higher than 15 kT. If the contribution from the magnetic attraction is considered, this value has to be much higher. From the experimental evaluation of the surface charge (zeta potential) of magnetic oxides, normally it is quite difficult to satisfy the above conditions, and obtaining stable suspension by controlling the solution composition using inorganic electrolytes is considered impossible. However, under the presence of both the surface adsorbed layer and electric double layer, the repulsive forces acting between particles can be represented by both equations (7.4) and (7.5), and consequently stable dispersion can be obtained relatively easily.

The total interaction force between particles U_T as a function of inter-particle distance is shown in figure 7.5 [66]. When r/D is a little larger than 1, the van der Waals attraction force becomes dominant. However, with the increase in inter-particle distance, steric repulsion becomes dominant in the case of a particle with surface adsorbed layers and Coulomb forces become dominant in the case of ion absorbed particles. Further increase in the inter-particle distance will lead to

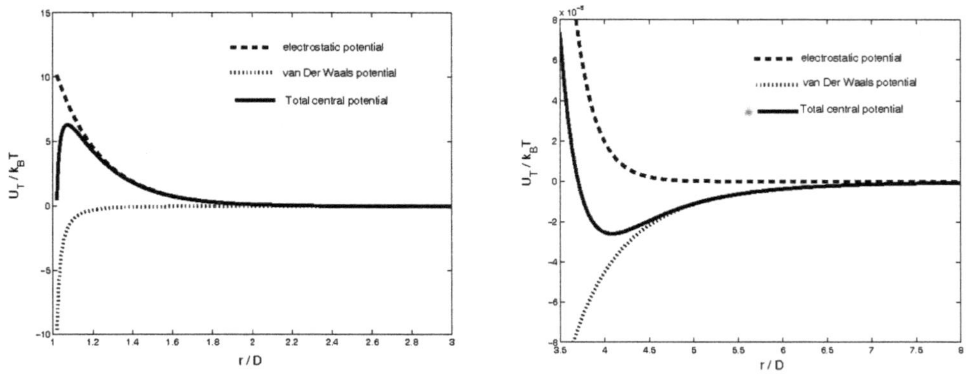

Figure 7.5. Inter-particle (a) central potential and (b) centered around the secondary minimum under an amplified energy scale, in units of kT at room temperature. The surface potential is assumed to be 60 mV [66].

instability due to the formation of the secondary minimum in the total potential curve.

7.4.2 Dispersion mechanism

To obtain a particle dispersion, it is not necessary to have a contribution from all the factors discussed in the previous section. In the case of the water-based magnetic fluid, to prevent the agglomeration of particles, besides the use of steric repulsion from surfactants, particle dispersion using the electric repulsion force using electrolytes also have been developed [64]. As a result, in the case of water-based magnetic fluid, the magnetic fluid is classified based on the origin of the repulsive force that is essential for the stable dispersion. In other words, magnetic fluids are prepared by using magnetic nanoparticles whose dispersion is achieved through (a) steric repulsion caused by the surfactant adsorbed on their surface and (b) electric repulsion due to the adsorption of polarized ions on the particle surface.

In the case of particle dispersion obtained using surfactant, stable suspension of the particles in the solvent is achieved by the inhibition of aggregation through the steric repulsion between particles generated by the surfactant adsorbed on their surface. Though this technique is used for the preparation of non-polar solvent-based magnetic fluid, this can be extended to the preparation of polar-based magnetic fluid such as water-based magnetic fluid. In other words, the dispersion of magnetic particles in a polar solvent is achieved by dispersing the monolayer surfactant coated hydrophobic magnetic particles through the introduction of another water-soluble surfactant. In this case, an anion or non-anion type surfactant having a high hydrophilic–hydrophobic balance (HLB) is used. Here, the surfactant introduced into water orients its hydrophobic end towards the surface of the magnetic particle and adsorbs physically with the hydrophobic end of the surfactant that is already chemically adsorbed on the surface of the magnetic particle. Consequently, the hydrophilic end of the surfactant becomes oriented towards the solvent and the affinity of the magnetic particle between the surface and water is enhanced. In addition, in the case of ionic surfactants, the surface charge inhibits the aggregation further and enhances the stability.

On the other hand, though the preparation of stable suspension of magnetic particles in aqueous media by controlling the surface charge alone is considered difficult, Massart *et al* [64] succeeded in obtaining a stable suspension by selecting the acid or base coordinating ion during the control over the potential determining ion H^+ and OH^- of the surface of oxide particles. Furthermore, the stable aqueous dispersion in acidic, neutral and alkaline regions has been developed by selecting appropriate potential determining ions. This has demonstrated that the dispersion of magnetic nanoparticles is possible even through the electric double layer effect and need not be restricted to the steric repulsion force alone. A schematic representation of magnetic particles' dispersions obtained utilizing the above-mentioned techniques is shown in figure 7.6.

The selection of the base liquid for the preparation of magnetic fluid depends on the application. Organic solvents are used mostly when engineering applications are

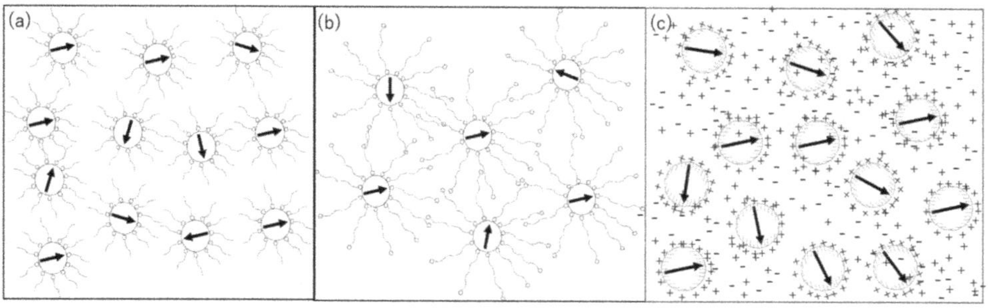

Figure 7.6. Schematic representation of magnetic particles' dispersions. (a) Dispersion in organic solvent through monolayer adsorption of surfactant, (b) dispersion in water through double layer adsorption of surfactant, (c) dispersion in water through the repulsive force from the electrical double layer.

considered. In these magnetic fluids, the magnetic particles are dispersed by the steric repulsion force generated by the surfactant adsorbed on the particle surface. Several other techniques have been developed to prepare magnetic colloids for biomedical applications [67].

7.5 Preparation of magnetic fluid

Magnetic fluid is a colloidal suspension dispersing fine ferromagnetic particles in either polar or non-polar solvent and possess both magnetic and fluidic properties. It is composed of ferromagnetic particles as the dispersed phase, the surfactant coats the particles and the solvent disperses the particles. The properties of magnetic fluid are influenced very much by the properties of the dispersed phase and the dispersing medium. Though the dispersing medium is decided based on the application, the physical properties of the magnetic fluid is determined by the magnetic properties of the dispersing phase. Aiming for the development of high quality magnetic fluid, magnetic oxide, metal and alloy dispersants with different magnetic properties have been proposed and developed. Here, the preparation of magnetic oxide and metal nanoparticle dispersed magnetic fluid is reported.

7.5.1 Magnetic oxide dispersed magnetic fluid

When the dispersant phase is composed of ferromagnetic oxides such as Fe_3O_4, $CoFe_2O_4$, (Mn-, Zn)Fe_2O_4 etc, the magnetic fluid is called an oxide-based magnetic fluid [68]. Depending on the application, water, kerosene, diester, alkylnaphthaline, fluorocarbon oil, etc, are used as the dispersing medium. However, with the advancement of the application technology, many new types of oxide-based magnetic fluid are already being sold utilizing the solvent properties. The main steps involved in the preparation of magnetic fluids are the synthesis of the magnetic oxide nanoparticles and their dispersion in various solvents. Here, we discuss in detail the preparation of magnetic fluid using magnetite as the dispersoid.

There are a number of techniques to synthesize magnetite nanoparticles, however, the co-precipitation technique that is generally used is introduced here [69]. In this method, first, solutions prepared dissolving 1 mol l^{-1} of ferrous chloride hydrate in

500 ml and 2 mol l^{-1} of ferric chloride hydrate in 500 ml are mixed. Then, this mixture is introduced into an alkaline solution prepared dissolving 6 N of sodium hydroxide (NaOH) in 1 l and the pH of the mixture is tuned to 11.5 ± 0.5 using NaOH to obtain magnetite nanoparticles. The chemical reaction for the formation of magnetite is given by the following equation.

$$FeCl_2 + FeCl_3 + 8NaOH \rightarrow Fe_3O_4 + 8NaCl + 4H_2O \quad (7.6)$$

The particle size can be controlled by varying the reaction temperature. The synthesized particles are recovered by adjusting the pH of the suspension to a value of 6.5, close to the isoelectric point of magnetite. This facilitates the aggregation of the particle that subsequently settles at the bottom of the vessel. The settled particles become suitable for the preparation of magnetic fluid after they are washed using distilled water a number of times to remove the dissolved ions.

The medium to disperse the particles obtained above is decided depending on the mechanism used to prevent the aggregation between the particles. The aggregation of the particles can be inhibited either by the steric repulsion generated by coating the particles with surfactant or by the electric double layer interaction induced by controlling the concentration of the surface charge determining ions in the medium as shown in figure 7.6.

7.5.1.1 Preparation of magnetic fluid using the steric repulsion induced by the adsorbed surfactant

The conditions for obtaining a stable dispersion through the adsorption of the surfactant on the surface of magnetic particles and the subsequent emergence of steric repulsion between particles can be summarized as follows:

(a) the functional group in the surfactant should irreversibly adsorb strongly on the surface of the particles.
(b) the chain length of the tail should possess the required length to maintain adequate inter-particle distance to overcome the van der Waals attraction.
(c) good intimacy between surfactant and solvent.

The preparation scheme to obtain surfactant adsorbed magnetite particles is as follows. First, an adequate amount of unsaturated fatty acid metal salt such as sodium oleate is introduced into the aqueous suspension dispersing co-precipitation derived magnetite for the formation of more than a monolayer adsorption on their surface. Then, the suspension is heated to 90 degrees to accelerate the adsorption of the ions of fatty acid on the surface of the particles. Then, the surfactant adsorbed particles are aggregated by adjusting the pH of the suspension between 5 and 7, filtered and washed to obtain the magnetic dispersoid. In the case of non-polar organic solvents, the particles obtained above are dried at 60 degrees for 48 h to remove the moisture prior to dispersing the same in the medium. On the other hand, in the case of preparing the magnetic fluid in a polar solvent such as water, water-soluble surfactant is introduced into the filtered cake of the monolayer surfactant adsorbed hydrophobic particles.

7.5.1.2 Preparation of magnetic fluid using the electric double layer interaction
The magnetic particles in the magnetic fluid prepared using the electric double layer interaction is stabilized even in the absence of any surfactant solely by the electric repulsion derived from the identical charge on the surface of the particles. The surface charge of magnetite particles dispersed in water becomes zero around pH 6.5 (pzc) and become negatively and positively charged at pH values, higher and lower than pzc, respectively. To neutralize the surface charge in each region, oppositely charged ions distribute in the vicinity of the surface. If the chemical action force with the ion and the surface hydration layer is extremely small, the behavior is chiefly dominated by the electric interaction and the resultant charge on the surface will be equivalent to the surface potential. Consequently, the electric repulsion energy, V_R becomes large. Using the method proposed by Massart *et al*, the particle dispersion in the alkali, acidic and neutral region were obtained by using tetramethyl ammonium [$N(CH_3)_4OH$], perchloric acid [$CHlO_4$] and citric acid [$C_6H_8O_7$] as potential determining ions. For example, in the case of obtaining magnetite suspension in the alkali region, first the magnetic particles are dispersed in distilled water and then tetramethyl ammonium hydroxide (TMAOH) is introduced. Here, OH^- ions adsorb on the surface of magnetite particles as surface potential determining ions and $[N(CH_3)_4]^+$ ions coordinate as ions with an opposite sign to facilitate the stable dispersion. For the successful preparation of magnetite particle dispersion in practice, the synthesized magnetite particle surface is made to become positively charged by maintaining the dispersion in acidic conditions. This will facilitate easy removal of the positively charged Na^+ ions that hinder the dispersion. This is achieved by washing the magnetite particle suspension with 0.1 N H_2SO_4 aqueous solution. Then, the pH of the suspension is tuned to 6.5 using 15% TMAOH and washed repeatedly prior to suction filtration to obtain magnetite cake with a moisture content of 70%. Finally, a specific amount of TMAOH is introduced to disperse the magnetite particles. The stable magnetic fluid under the alkali region is obtained after subjecting the dispersion to centrifugal separation to remove the agglomerates. Similarly, the stable suspension in the acidic region can also be obtained. However, when citric ion is used as the potential determining ion, besides the electric repulsion force, the steric repulsion force generated due to the adsorption of citric acid molecules on the surface of the particles will also be added and the suspension becomes more stable.

7.5.2 Metallic magnetic fluid

A report on the preparation of metallic magnetic fluid dispersing metallic or alloy nanoparticles appeared soon after the invention of magnetic fluid. The strong interest was due to the fact that the magnetizations of these particles are a couple or more times higher than their oxide counterparts and also these could be dispersed in a medium that possesses high thermal and electrical conductivities [70, 71]. However, outstanding progress was recorded only in the middle of the 1980s.

In Japan, though a number of reports from Nakatani and Wakayama were published [72–74], the iron nitride magnetic fluid developed by Nakatani *et al* was the first of its kind [75]. However, iron nitride was highly unstable in an oxidizing atmosphere. Though the development of an oxidation protection technique was actively pursued, no considerable progress was reported. In 2002, Bonnemann *et al* [76] worked on the synthesis of metallic magnetic particles by decomposing metallic carbonyl in tetrahydronaphthalene and succeeded in the preparation of stable magnetic fluid dispersing iron and iron–cobalt nanoparticles [77]. The preparation as well as the circumstances that led to the development of the technique to obtain stable metallic nanoparticles in an oxidizing atmosphere is described below.

7.5.2.1 Preparation of magnetic fluid dispersing iron–cobalt alloy nanoparticles
The basic steps involved in the preparation of iron nanoparticles dispersed magnetic fluid were as follows: 25.52 mM of $Al(C_8H_{17})_3$ was dissolved in 500 ml of tetrahydronaphthalene and 255.2 mM of metallic carbonyl was introduced. Then, the system was stirred and heated gently under argon atmosphere up to 265 °C and

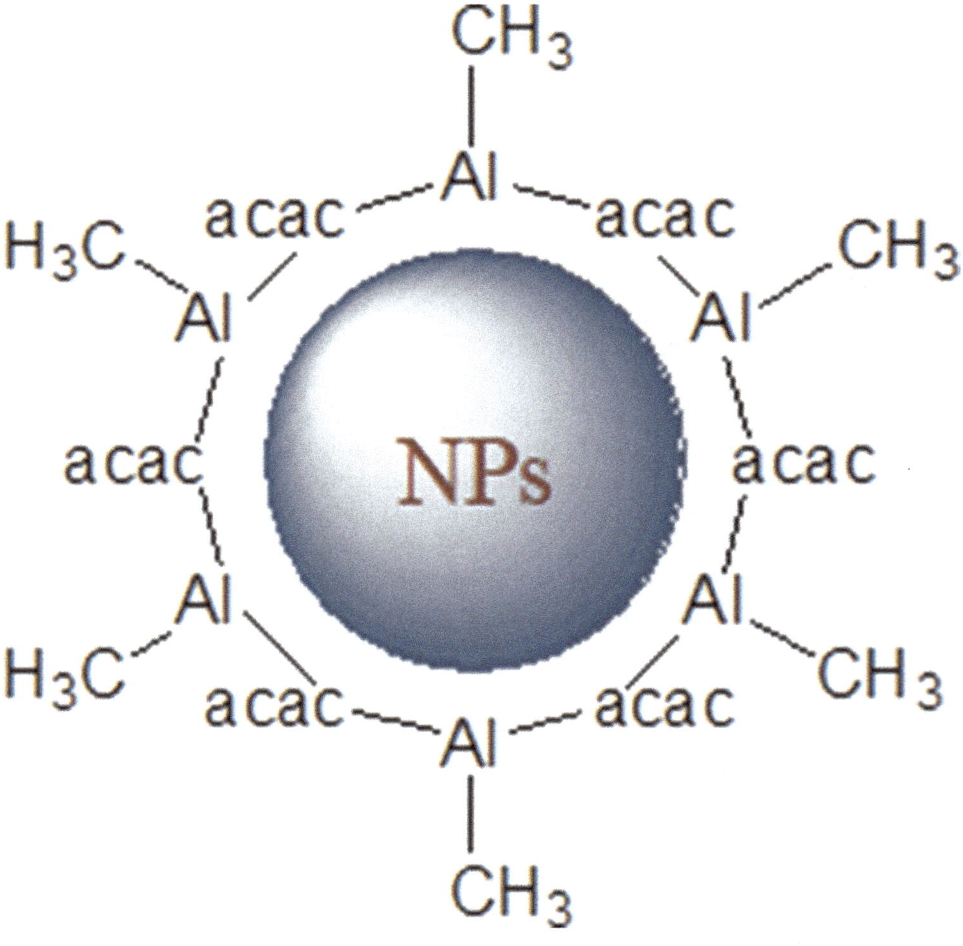

Figure 7.7. $Al(C_8H_{17})_3$ coated metallic and alloy nanoparticles synthesized by the thermal decomposition technique.

maintained at this temperature for an hour. Then, the system was heated up to 313 °C at the rate of 10 degrees per minute and maintained at this temperature for 5 h to obtain $Al(C_8H_{17})_3$ coated iron nanoparticles (figure 7.7). Then, the system was cooled to room temperature and purged with Ar (3.5% O_2) gas for three hours prior to washing the particles with toluene and storing the same in toluene. Immediately after this step, either anacardium occidentale or courantyne was introduced. Finally, the magnetic fluid was prepared by dispersing the above particles either in kerosene or toluene.

In the case of iron–cobalt magnetic fluid, besides the preparation of the iron–cobalt carbonyl precursor, the preparation technique is similar to the one described above for iron magnetic fluid. Iron–cobalt carbonyl precursor was prepared by dissolving 255.2 mM of iron carbonyl and 63.8 mM of cobalt carbonyl in 500 ml of tetrahydronaphtalene and stirring the same for 72 h at room temperature. For details, refer to the article published by Bonnemann *et al* [77].

In this section, the preparation of magnetic fluid dispersing metallic oxide and metal/alloy nanoparticles was described. The properties of each magnetic fluid are given in table 7.1. Besides the above magnetic fluids, magnetic fluids that utilize the properties of magnetic particles such as temperature sensitivity and the properties of the medium such as viscosity and low vapor pressure have also been developed to meet the needs of industrial applications [78–80].

7.6 Magnetic fluid based applications

7.6.1 Engineering applications

The properties of magnetic fluid have been utilized in the development of speakers, spindle motors, hard disc drives, vacuum seals, printers, oil–water separation, visualization of magnetic patterns in magnetic recording media, etc, in engineering applications [81, 82]. When an external magnetic field is applied onto magnetic fluid, the apparent density of the magnetic fluid increases due to the magnetic body force generated depending on the magnetic field gradient and the product of magnetic volume and magnetization. The degree of increase is a function of the concentration of ferromagnetic particles, the magnitude of the magnetic field and field gradient and an apparent density as high as 10 is practically possible. Using the above property, engineering application devices for density separation of non-magnetic materials, polishing the surface of a non-magnetic body, have been developed [83, 84].

However, when a magnet is introduced into magnetic fluid placed inside a non-magnetic container, the magnet will float almost in the middle of the container due to the magnetic body force generated. Utilizing the above phenomenon, development of acceleration sensors, inclination sensors, dampers, etc, has also been considered. Another important phenomenon observed in magnetic fluid is the magnetic convection generated due to the magnetization-temperature dependence of magnetic particles and induction in magnetic field. Energy conversion devices are being developed based on the above phenomenon [85, 86]. As already described here, the application of magnetic fluid in multiple-diverse fields has been considered

Magnetic fluid	Color of the liquid	M_s(25 °C	Viscosity(mPa s)	Solvent	Application
W-35	Black	360	30 (25 °C)	Water	Density separation
HC-50	Blackish brown	420	30 (25 °C)	Kerosene	Density separation
DEA-40	Black	400	400 (25 °C)	Diester	Seal, damper, bearing
NS-35	Black	350	900 (25 °C)	Alkyl Naphthalene	Vacuum seal, damper, bearing
PX-10	Black	100	1000 (25 °C)	Mineral oil	Damper
CFF 200 A	Shiny black–blackish brown	250	300 (27 °C)	Ester	Conductive dust proof, seal
VSG 600 series	Shiny black–blackish brown	200–600	110–2100 (27 °C)	Ester/hydrocarbon	Vacuum seal
APG 800, 900 REN series	Shiny black–blackish brown	100–350	100–6000 (27 °C)	Poly alpha olefin	Speaker
P series	Shiny black –blackish brown	≦100	≦5 (27 °C)	Water/hydrocarbon	Observation of magnetic pattern
Fe_3N[38]	Black	1700–2400		Kerosene	Seal, damper, speaker
Fe–Co[40)]		About 840		Kerosene	Seal, bearing, speaker

and become possible. In the next section, we discuss the technology developed based on magnetic particles and magnetic fluid.

7.6.2 Biomedical applications

7.6.2.1 Molecular detection and control using MNPs

The immobilization of biomolecular entities using MNPs is being attempted to (a) control the localization or isolation of specific biological targets by applying an external magnetic field and (b) dephasing the spin–spin relaxation times of the proton in the surrounding media exploiting the superparamagnetic behavior. Here, the degree of magnetic interference is a function of the MNP assembly caused by their interaction with the surroundings. Consequently, many applications using bio-functionalized MNPs such as molecular bio-sensing [87–92], DNA separation [93–98], and pathogen detection and sequestration [99–102] have been explored.

7.6.2.2 MRI contrast agent

In vivo molecular imaging techniques are used for studying diseases non-invasively at the molecular level to visualize the molecular characteristics of physiological or pathological processes in living organisms without invasive procedures. MRI offers 3D images with excellent spatial resolution with optimal contrast and a very high signal-to-noise ratio. With respect to molecular imaging, iron oxide based MNPs such as magnetite/maghemite that induce a stronger contrast have been favored and are being practically used [103].

7.6.2.3 Drug delivery system

The use of biocompatible and high specificity magnetic oxide nanoparticles has been considered for biological targeting exploiting magnetic guidance using the magnetic field, to increase the accumulation of drugs to diseased sites for over nearly half a century. Several improved MNP models have been developed, particularly for cancer therapy. However, despite very promising *in vitro* results, the *in vivo* studies have not been found to be very effective and clinically approved magnetic nano-carriers are yet to be commercialized [104, 105].

Targeting of MNPs using an external magnetic field is expected to bring about a rise in the local concentration of the administered drug and could lead to a reduction in overall dose. However, the method is challenging due to the nanosize of MNP in spite of the fact that this technique could give the MNPs more chance to be trapped by the magnetic field close to the target area. A single Fe_3O_4 MNP with a core diameter of less than 20 nm cannot be captured permanently even by a high-gradient magnetic separation column. Thus, the use of phospholipid-coated or polymer-coated MNP agglomerates has been considered [106]. The movement of MNPs inside a matrix or fluid depends directly on a multitude of factors including the interaction between MNPs and fluid components, size and shape, among others. *In vivo* studies related to dynamics transport of MNPs through a vein or artery to a targeted area are being carried out [107, 108].

On the other hand, active targeting that could improve the accumulation of chemotherapeutics at the tumor site is also being considered. A major barrier to the advancement of this technique is that this requires stable suspension of MNPs with tailored chemical properties [109–111].

Considerable progress has been reported using MNPs for actively targeted drug delivery. It is claimed that the 'enhanced permeation and retention' (EPR) due to a passive selective delivery process as a consequence of diseased vascular condition, facilitates the release of drugs from passively diffusing PEGylated nanocarriers through peripheral tumor tissue. This process depends on the biophysical properties of the MNPs [112–114].

7.6.2.4 Magnetic hyperthermia

Hyperthermia itself is an old concept, however, it has received renewed attention due to the recent advances in the synthesis of MNPs that act as thermal seeds. This technique could become a potential application in cancer therapy [115]. The heat generated by the MNPs exposed to an AC magnetic field annihilates cancer cells while keeping the normal tissue unharmed. Heating is obtained through either Brownian or Néel relaxation of the magnetic vector within the MNPs subjected to an AC magnetic field [116]. The efficiency of magnetic heating essentially depends on the physical properties of the MNPs such as size and magnetic susceptibility [117]. The development in the synthesis technique of MNPs with varying magnetic properties through finer control over their size, shape and structure has increased the possibility of treating cancer using magnetic hyperthermia in the near future. Consequently, whole-body magnetic hyperthermia, a carcinogenicity-free ideal cancer therapy that utilizes the lowered heat resistance of cancerous tissues that undergo cell death at temperatures within the range of 42–43 °C using magnetic nanoparticles as thermal seeds that enables selective heating of undetectable hidden micro tumors in internal body parts, has been proposed and intensively researched [118–120].

In the recent past, considerable research has been carried out to determine the potential of this proposal focusing mostly on the heat generation capacity of the thermal seeds under practical conditions using the radio-waveband AC magnetic field. The size of the thermal seeds that enable safe delivery to cancers hidden deep inside the body is considered to fall between 8 and 25 nm considering the practical issues such as drug delivery techniques, gaps in blood vessel walls, stable dispersion criteria, etc. Besides the physical properties of the magnetic seeds, delivering the same in appropriate quantities to targeted sites is another vital issue to be considered for effective therapy. However, the accumulation of iron oxide nanoparticles with an average core and hydrodynamic diameters 11.3 nm and 23.8 nm, respectively, in the tumor has been evaluated to be only 1.9 mgFe cm^{-3} [121]. Considering the fact that the cancerous areas should be heated up to 50 °C for complete cell annihilation [122], for effective thermotherapy, heat generation from such an extremely small amount of thermal seeds under an allowable AC field and frequency conditions have to be maximized within the framework of the physics of nanomagnetism.

Besides eddy current loss, though the conversion of magnetic energy to heat energy is considered to be due to hysteresis loss, relaxation loss and friction loss are also losses generated from the same source and could be made to generate considerable heat by tuning the field strength and frequency conditions to match specific loss type. Consequently, the heating power generated in magnetic nanoparticles, P_h becomes limited to $P_{hMAX} = 4\rho^{-1}\mu_0 M_s H_{ac} f$ regardless of the type of heat generation mechanism. However, the solution to enhance P_h toward P_{hMAX} differs according to the kind of mechanism.

If superparamagnetic nanoparticles are considered as thermal seeds, the efficiency of heat generation ε (= P_h/P_{hMAX}) of relaxation loss can be maximized when $2\pi f$ is adjusted to $\tau^{-1} = \tau_N^{-1} + \tau_B^{-1}$ and H_{ac} is increased as to be high as possible within the linear response region. Apparently, both the efficiency ε and maximum value P_{hMAX} of heat generation increase if the amplitude H_{ac} of the AC magnetic field is increased. However, it should be noted that several constraints such as human safety [123], design of the AC magnetic field applicator, etc, on such a strategy exists. But, unfortunately most of the research investigations are being carried out without giving due considerations to the above facts. A relatively recent simulation study on the electromagnetic field distribution and the flow of heat arising from irradiation using a detailed human body model, looking for the answer to safe field and frequency conditions for magnetic particle imaging [124], have shown that the allowable upper limit of $H_{ac}f$ for irradiation on the whole human body should be roughly 2×10^8 $Am^{-1}s^{-1}$ to avoid the generation of hot spots deep inside parts of the human body. Considering the above fact along with the concentration of magnetic particles in targeted sites, it is obvious that the heat generated would not be adequate enough for cancer treatment. Thus, besides the physical properties of the magnetic seeds, the accumulation of magnetic thermal seed concentration should be increased ten-fold using advanced drug delivery technology to enhance the nanoparticle uptake.

In addition, a practical solution for effective whole-body magnetic fluid hyperthermia treatment under the present state of magnetic particle uptake considering magnetic particles with enhanced saturation magnetization and the magnetic field application technique has been proposed [125]. In the context of limited success of the conventional hyperthermia treatment using magnetite nanoparticles with respect to tiny cancers scattered throughout the whole body, the successful development of a core–shell structured particle composed of an iron core and biocompatible iron oxide shell with a saturation magnetization $\mu_0 M_s$ of 1.2 T at 14 nm [126] and 1.6 T at 40 nm [127] has been reported. The use of such particles will lead to the enhancement in unit weight saturation magnetization $\rho^{-1}\mu_0 M_s$, which is one of the factors that would lead to higher thermal energy.

Also, the upper limit of H_{ac} and f that contribute to enhanced heat efficiency is also reconsidered from the view point of hot spot generation conditions reported by Dossel *et al* [124]. The potential of heating by pulsed irradiation using the above-mentioned core–shell nanoparticles and allowing the use of quadrupled power at $H_{ac}f$ of 8×10^8 Am^{-1} s^{-1} under a feasible nanoparticles concentration of 1.9 mgFe cm^{-3} in a tumor [121] has been considered. The simulations have shown that the temperature for ε of 0.8 would exceed 50 °C for an exposure time corresponding to $t = 60$ s, while the temperature for ε of 0.4 would never reach 45 °C. An alternative technique for whole-body treatment limiting the exposure time to 180 s for ε, 0.8 has been proposed (figure 7.8). This exemplified calculation has showed that tiny cancers of roughly 1 cm could be eradicated without serious side effects on the hot spot portion, by performing a short pulse irradiation to the newly selected core–shell structured nanoparticles. Thus, the whole-body treatment with the ingenious schemes unemployed in conventional hyperthermia treatments can remain as a prospective option for cancer therapy if the efficiency ε is considerably improved in comparison with the currently achieved level ($\varepsilon < 0.4$). This would also lead to research in the development of core–shell magnetic nanoparticles with enhanced properties.

In addition, the possibility of improving ε to a practical level has been corroborated by gathering knowledge about the magnetic properties of nanoparticles beyond the understanding based on the conventional models and is necessary to improve the design of magnetic nanoparticles for raising ε to the utmost limits. The non-linear and non-equilibrium responses of thermally fluctuating systems for magnetic particles under different relaxation modes and field

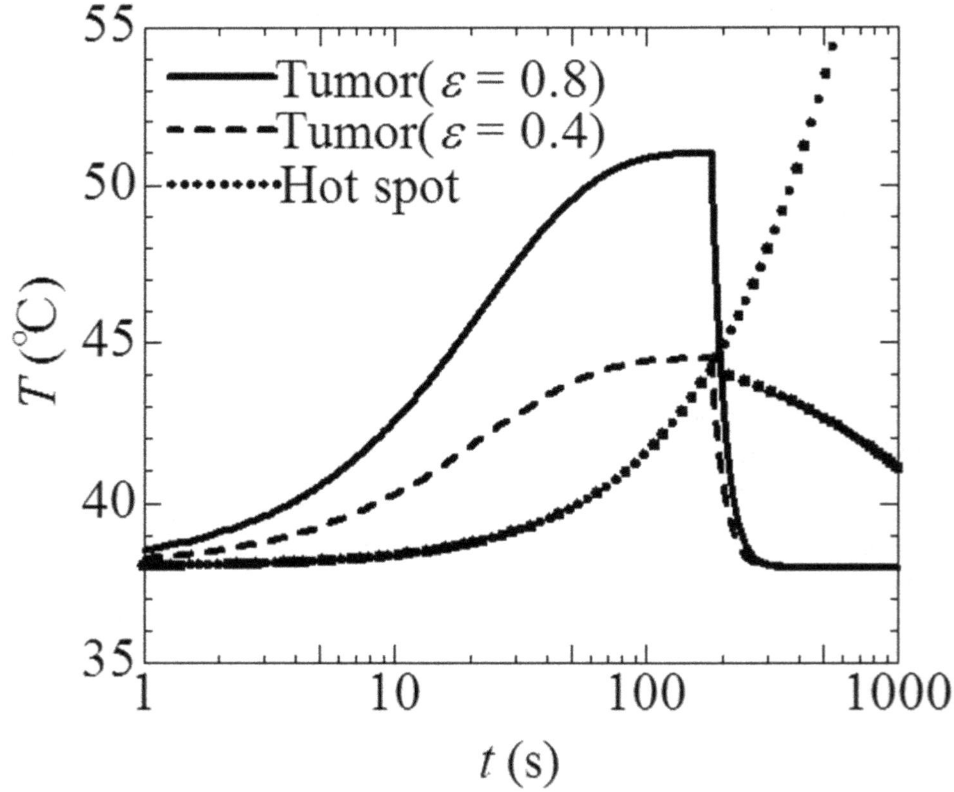

Figure 7.8. Calculated temperature elevation curves of tiny tumors and the hot spot portion in normal muscle tissue by pulsed irradiation of the AC magnetic field.

conditions such as (a) Néel relaxation in magnetic fields, (b) Brownian relaxation in magnetic fields, (c) easy axes oriented toward directions parallel, or perpendicular to the AC field, (d) magnetic hysteresis of superparamagnetic states, (e) intermediate state between ferromagnetic and superparamagnetic nanoparticles have been studied using numerical simulations [128, 129]. The results have indicated that the non-equilibrium structure formed with the orientation of the easy axes in large AC magnetic fields could increase ε to almost 0.8 under certain conditions. It could be anticipated that the results of the numerical simulation will be experimentally verified in the near future.

7.7 Summary

The necessary technology for the preparation of magnetic particles and their subsequent dispersion in aqueous and non-aqueous solvents for the preparation of magnetic fluid/suspension to be used in the field of engineering and biomedical applications has already been well established. However, the importance of the development of particles with advanced magnetic properties for enhanced performance of already existing applications and also for further expansion in applied technology is well recognized. Consequently, considerable efforts are being made to promote magnetic fluid research, and subsequently the successful development of magnetic particles such as metallic iron and iron–cobalt alloy particles using advanced synthesis techniques has been realized. Though the size as well as the stability of such particles in atmospheric conditions was considered to limit their use, the research and development of the technology proposed in the recent past for the synthesis of core–shell structured iron oxide stabilized metallic iron particles should be given due consideration. Any advancement in the synthesis of nanosized core–shell structured iron–iron oxide magnetic particles could become a potential solution for their use in both engineering and biomedical applications.

References

[1] Nakatani I Iron nitride magnetic fluid, personal communication.

[2] Papell S S 1965 Low viscosity magnetic fluid obtained by the colloidal suspension of magnetite particles. US Patent No.3,215,572.

[3] Satoh T, Higuchi S and Shimoiizaka J 1966 Dispersion property of a magnetite colloid in cyclohexane and van der Waals energy 19th *Annu. Meeting of the Chemical Society of Japan* p 293.

[4] Shimoiizaka J 1976 Patent 51-44580

[5] Shimoiizaka J 1976 Patent 53-17118

[6] Reimers G W and Khalafalla S E 1974 Production of magnetic fluids by peptization techniques *US Patent No.3* **843** 540

[7] Sato T 1970 Formation and magnetic properties of ultrafine spinel ferrites *IEEE Trans. Magn.* **6** 795–9

[8] Sato T, Iijima T, Seki M and Inagaki N 1987 Magnetic properties of ultrafine ferrite particles *J. Magn. Magn. Mater.* **65** 252–6

[9] Jeyadevan B, Chinnasamy C G P N, Perales Perez O, Shinoda K, Tohji K and Kazuya A 2003 Auxiliary aqueous solution processes for the synthesis of cobalt ferrite with high coercivity at room temperature *J. Jpn. Soc. Powder Powder Metall.* **50** 114–9
[10] Chinnasamy C N, Senoue M, Jeyadevan B, Perales-Perez O, Shinoda K and Tohji K 2003 Synthesis of size-controlled cobalt ferrite particles with high coercivity and squareness ratio *J. Colloid Interface Sci.* **263** 80–3
[11] Chinnasamy C N, Jeyadevan B, Shinoda K, Tohji K, Djayaprawira D J, Takahashi M, Joseyphus T J and Narayanasamy A 2003 Unusually high coercivity and critical single-domain size of nearly monodispersed $CoFe_2O_4$ nanoparticles *Appl. Phys. Lett.* **83** 2862
[12] Brinker C J and Sherrer G W 1990 *Sol-gel Science: The Physics and Chemistry of Sol-gel Processing* (New York: Academic)
[13] Lee J-G, Lee H M, Kim C S and Oh Y-J 1998 Magnetic properties of $CoFe_2O_4$ powders and thin films grown by a sol–gel method *J. Magn. Magn. Mater.* **177** 900–2
[14] Chen D H and He X R 2001 Synthesis of nickel ferrite nanoparticles by sol–gel method *Mater. Res. Bull.* **36** 1369–77
[15] Duxin N, Stephan O, Petit C, Bonville P, Colliex C and Pileni M P 1997 Pure α-Fe coated by an $Fe_{1-x}B_x$ alloy *Chem. Mater.* **9** 2096–100
[16] Tanori J, Duxin N, Petit C, Lisiecki I, Veillet P and Pileni M P 1995 Synthesis of nanosize metallic and alloyed particles in ordered phases *Colloid Polym. Sci.* **273** 886–92
[17] Yener D O and Giesche H 2001 Synthesis of pure and manganese-, nickel-, and zinc-doped ferrite particles in water-in-oil microemulsions *J. Am. Ceram. Soc.* **84** 1987–95
[18] Wilcoxon J P and Provencio P P 1999 Use of surfactant micelles to control the structural phase of nanosize iron clusters *J. Phys. Chem.* B **103** 9809–12
[19] Naito S 2002 *Catal. Soc. Jpn.* **44** 253
[20] Jia Q, Zeng J, Qiao R, Jing L, Peng L, Gu F and Gao M 2011 *J. Am. Chem. Soc.* **133** 19512
[21] Sun S H and Zeng H 2002 *J. Am. Chem. Soc.* **124** 8204
[22] Gao J, Liang G, Cheung J S, Pan Y, Kuang Y, Zhao F, Zhang B, Zhang X, Wu E X and Xu B 2008 *J. Am. Chem. Soc.* 13011828
[23] Ho C-H, Tsai C-P, Chung C-C, Tsai C-Y, Chen F-R, Lin H-J and Lai C-H 2011 *Chem. Mater.* **23** 1753
[24] Lynch J, Zhuang J, Wang T, LaMontagne D, Wu H and Cao Y C 2011 *J. Am. Chem. Soc.* **133** 12664
[25] Palchoudhury S, An W, Xu Y, Qin Y, Zhang Z, Chopra N, Holler R A, Turner C H and Bao Y 2011 *Nano Lett.* **11** 1141
[26] Park J, An K, Hwang Y, Park J-G, Noh H-J, Kim J-Y, Park J-H, Hwang N-M and Hyeon T 2004 *Nat. Mater.* **3** 891
[27] Bao N, Shen L, An W, Padhan P, Heath Turner C and Gupta A 2009 *Chem. Mater.* **21** 3458
[28] Bau J A, Li P, Marenco A J, Trudel S, Olsen B C, Luber E J and Buriak J M 2014 *Chem. Mater.* **26** 4796
[29] Fukumoto H and Jeyadevan B 2017 unpublished data
[30] Ely T O, Amiens C and Chaudret B 1999 Synthesis of nickel nanoparticles. Influence of aggrehation induced by modification of poly(vinylpyrrolidone) chain length on their magnetic properties *Chem. Mater* **11** 526–9
[31] Hyeon T, Lee S-S, Park J, Chung Y and Na H B 2001 Synthesis of highly crystalline and monodisperse maghemite nanocrystallites without a size-selection process *J. Am. Chem. Soc.* **123** 12798–801

[32] Khomutov G B, Yu Obydenov A, Yakovenko S A, Soldatov E S, Trifonov A S, Khanin V V and Gubin S P 1999 Synthesis of nanoparticles in Langmuir monolayer *Mater. Sci. Eng.: C* **8–9** 309

[33] Kumar R V, Koltypin Y, Cohen Y S, Cohen Y, Aurbach D, Palchik O O, Felner I and Gedanken A 2000 Preparation of amorphous magnetite nanoparticles embedded in polyvinyl alcohol using ultrasound radiation *J. Mater. Chem.* **10** 1125–9

[34] Puntes V F, Krishnan K M and Alivisatos A P 1999 Colloidal nanocrystal shape and size control: The case of cobalt *Science* **291** 2115–7

[35] Atsumi T *et al* 2005–12 *Japan Magnetic Fluid Society Meeting* 36

[36] Fujita T, Jeyadevan B and Mamiya M 1989 Effect of surfactants on the size of iron particles formed by the reduction of ferrous ion in aqueous solution *J. Jpn. Soc. Powder Powder Metall* **36** 778–81

[37] Lu J, Dreisinger D B and Cooper W C 1997 Cobalt precipitation by reduction with sodium borohydride *Hydrometallurgy* **45** 305–10

[38] Glavee G N, Klabunde K J, Sorensen C M and Hadjipanayis G C 1994 Borohydride reduction of nickel and copper ions in aqueous and nonaqueous media: Controllable chemistry leading to nanoscale metal and metal boride particles *Langmuir* **10** 4726

[39] Yedra A, Fernandez Barquin L, Garcia Calderon R, Pankhurst Q A and Gomez Sal J C 2001 Survey of conditions to produce metal–boron amorphous and nanocrystalline alloys by chemical reduction *J. Non-Cryst. Solids* **287** 20–5

[40] Fiévet F, Figlarz M and Lagier J-P 1983 Process for reducing metallic compounds using polyols, and metallic powders produced thereby. Europe Patent 0113,281

[41] Viau G, Fiévet-Vincent F and Fiévet F 1996 Monodisperse iron-based particles: Precipitation in liquid polyols *J. Mater. Chem.* **6** 1047–53

[42] Silvert P-Y, Vijayakrishnan V, Vibert P, Herrera-Urbina R and Elhsissen K T 1996 Synthesis and characterization of nanoscale Ag-Pd alloy particles *Nanostruct. Mater.* **7** 611–8

[43] Viau G, Fiévet-Vincent F and Fiévet F 1996 Nucleation and growth of bimetallic CoNi and FeNi monodispersed particles prepared in polyols *Solid State Ionics* **84** 259–70

[44] Kurihara L K, Chow G M and Schoen P E 1995 Nanocrystalline metallic powders and films produced by the polyol method *Nanostruct. Mater.* **5** 607–13

[45] Chow G M, Kurihara L W, Kemner K M, Schoen P E, Elam W T, Ervin A, Keller S, Zhang Y D, Budnick J and Ambrose T 1995 Structural, morphological, and magnetic study of nanocrystalline cobalt-copper powders synthesized by the polyol process *J. Mater. Res.* **10** 1546–54

[46] Justin J R *et al* 2004 *Proc. of 28th Meeting of the Mag. Soc. of Japan*

[47] Perales O P *et al* 2001 *Proc. Int. Symp. on Cluster Assembled Matter, IPAP Conf. Series* vol 3, p. 105

[48] Jeyadevan B, Hobo A, Perales-Perez O, Chinnasamy C G P N, Shinoda K, Tohji K and Kasuya A 2003 Solution synthesis of nanocrystalline core–shell structured NiCo particles *J Jpn. Soc. Powder Powder Metall.* **50** 107–13

[49] Jeyadevan B, Uehara K, Hobo A, Chinnasamy N, Shinoda K, Tohji K, Djayaprawira D D J, Tsunoda M and Takahashi M 2003 Direct synthesis of fct-FePt nanoparticles by chemical route *Jpn. J. Appl. Phys.* **42** L350

[50] Sato K *et al* 2004 *Int. Conf. on Magnetic Fluid, Brazil 2–6 August*

[51] Chinnasamy C N, Jeyadevan B, Shinoda K and Tohji K 2003 Polyol-process-derived CoPt nanoparticles: Structural and magnetic properties *J. Appl. Phys.* **93** 7583

[52] Chinnasamy C N *et al* 2005 *Proc. of 28th Meeting of the Mag. Soc. of Japan*

[53] Chinnasamy C N *et al* 2004 *Proc. of 28th Meeting of the Mag. Soc. of Japan*

[54] Hinotsu T, Jeyadevan B, Chinnasamy C N, Shinoda K and Tohji K 2004 Size and structure control of magnetic nanoparticles by using a modified polyol process *J. Appl. Phys.* **95** 7477

[55] Matsumoto T, Takahashi K, Kitagishi K, Shinoda K, Cuya Huaman J L, Piquemal J-Y and Jeyadevan B 2015 Dissolution and reduction of cobalt ions in the polyol process using ethylene glycol: identification of the active species and its role *New J. Chem.* **39** 5008–18

[56] Takahashi K, Yokoyama S, Matsumoto T, Cuya H J L, Piquemal J-Y, Miyamura H and Jeyadevan B 2016 Towards designed synthesis of metallic nanoparticles-elucidation of the redox scheme in cobalt-ethylene glycol system *New J. Chem.* **40** 8632

[57] Kaneko H and Jeyadevan B unpublished data

[58] Willard M A, Kurihara L K, Carpenter E E, Calvin S, Calvin S and Harris V G 2004 *Encyclopedia of Nanoscience and Nanotechnology* vol 1 (Valencia, CA: American Scientific Publishers), 821

[59] Bönnemann H and Nagabhushana K S 2004 *Encyclopedia of Nanoscience and Nanotechnology* vol 1 (Valencia, CA: American Scientific Publishers), 777

[60] Dong H, Chen Y-C and Feldmann C 2015 Polyol synthesis of nanoparticles: status and options regarding metals, oxides, chalcogenides, and non-metal elements *Green Chem.* **7** 4107–32

[61] Kodama D, Jeyadevan B, Shinoda K, Sato K, Sato Y and Tohji K 2006 Synthesis of FeCo alloy particles by modified polyol process and their structural and magnetic properties 2006 *IEEE Int. Magnetics Conference* (*INTERMAG*)

[62] Rosensweig R E 1985 *Ferrohydrodynamics* (London: Cambridge University Press)

[63] Israelachvili J N 1991 *Intermolecular and Surface Forces* (London: Academic)

[64] Massart R 1980 Magnetic fluids and process for obtaining them, *US Patent No 4*,329,241

[65] Dubois E 1998 *PhD Thesis* Universite Pierre et Marie Curie, Paris

[66] Scherer C and Figueiredo Neto A M 2005 Ferrofluids: Properties and applications *Braz. J. Phys.* **35** 718–27

[67] Kikuchi T, Kasuya R, Endo S, Nakamura A, Takai T and NilsMetzler-Nolte 2011 Preparation of magnetite aqueous dispersion for magnetic fluid hyperthermia *J. Magn. Magn. Mater.* **323** 1216–22

[68] Nakatsuka K 1993 Magnetic Fluid Seminar (Magnetic Fluid Society of Japan) p 1

[69] Tsuda S 1985 *Master's Thesis* Tohoku University, Japan

[70] Thomas J R 1966 Dispersions of small discrete particles of ferromagnetic metals *US Patent No 3* **228** 881

[71] Berkowitz A E and Walter J L 1983 Ferrofluid *US Patent No 4* **381** 244

[72] Nakatani I *et al* 1985 Patent S60-162704; Wakayama K and Harada T 1986 *Patent* S61-36907

[73] Nakatani I and Furubayashi T 1987 *Patent* 62-11207

[74] Wakabayashi K *et al* 1988 *Patent* S63-164404

[75] Nakatani I and Furubayashi T 1994 *Patent* H2-04026

[76] Bönnemann H, Brijoux W, Brinkmann R, Matoussevitch N and Waldorfner N 2002 DE102227779.6 Studiengesellschaft Kohle mbH.

[77] Bönnemann H *et al* 2005 Air stable Fe and Fe-Co magnetic fluids-synthesis and characterization *Appl. Organomet. Chem.* **19** 790–6

[78] Nakatsuka K *et al* 1991 *Japanese Magnetic Fluid Society Meeting* 25–7

[79] Shimoiizaka J 1997 *Patent* 52-782

[80] Shimoiizaka J 1977 *Patent* 52-783

[81] Tsuda S 2002 *Nanoryuushikougakudaikei, Part II Application Technology* (Fuji Techno System Co Ltd), 257–64

[82] Fujita T 2004 *Frontier Technology of Nano and Micro particles* (JMC Publishing Co, Ltd), 292–8

[83] Shimoiizaka J 1983 *Koubutsugakuzasshi* **16** 149–56

[84] Umehara N, Raghunandan M and Komanduri R 1994 *Japanese Magnetic Fluid Society Meeting* 69–71

[85] Rosensweig R E, Nestor J W and Timmins R S 1965 Ferrohydrodynamics for direct conversion of heat energy *Proc. Symp. AICh-I. Chem. Eng. Ser.* **5** 104–18

[86] Nakatsuka K 1990 *Fine Particles Handbook* (Fuji Techno System Co Ltd), 592–6

[87] Perez J M, O'Loughin T, Simeone F J, Weissleder R and Josephson L 2002 *J. Am. Chem. Soc.* **124** 2856–7

[88] Perez J M, Josephson L, O'Loughlin T, Hogemann D and Weissleder R 2002 *Nat. Biotechnol.* **20** 816–20

[89] Taktak S, Sosnovik D, Cima M J, Weissleder R and Josephson L 2007 *Anal. Chem.* **79** 8863–9

[90] Perez J M, Josephson L and Weissleder R 2004 *ChemBioChem* **5** 261–4

[91] Koh I, Hong R, Weissleder R and Josephson L 2008 *Angew. Chem. Int. Ed.* **47** 4119–21

[92] Lee H, Yoon T J and Weissleder R 2009 *Angew. Chem., Int. Ed.* **48** 5657–60

[93] Obata K, Tajima H, Yohda M and Matsunaga T 2002 *Pharmacogenomics* **3** 697–708

[94] Safarik I and Safarikova M 2004 *Biomagn. Res. Technol.* **2** 7

[95] Gu H W, Xu K M, Xu C J and Xu B 2006 *Chem. Commun.* 941–9

[96] Uhlen M 1989 *Nature* **340** 733–4

[97] Hong J W, Studer V, Hang G, Anderson W F and Quake S R 2004 *Nat. Biotechnol.* **22** 435–9

[98] Tanaka T, Sakai R, Kobayashi R, Hatakeyama K and Matsunaga T 2009 *Langmuir* **25** 2956–61

[99] Ugelstad J, Kilaas L, Stenstad P, Ellingsen T, Bjorgum J, Aune O, Nilsen T N, Schmid R and Berge A 1991 (Wordsmiths' Conference Publications)

[100] Gu H, Ho P L, Tsang K W, Wang L and Xu B 2003 *J. Am. Chem. Soc.* **125** 15702–3

[101] Kell A J, Stewart G, Ryan S, Peytavi R, Boissinot M, Huletsky A, Bergeron M G and Simard B 2008 *ACS Nano* **2** 1777–88

[102] Gao J, Li L, Ho P L, Mak G C, Gu H and Xu B 2006 *Adv. Mater.* **18** 3145–8

[103] Bellin M F 2006 *Eur. J. Radiol.* **60** 314–23

[104] Prijic S and Sersa G 2011 *Radiol. Oncol.* **45** 1–16

[105] McBain S C, Yiu H H and Dobson J 2008 *Int. J. Nanomed.* **3** 169–80

[106] Moeser G D, Roach K A, Green W H and Hatton T A 2005 *AIChE J.* **50** 2835–48

[107] Nacev A, Beni C, Bruno O and Shapiro B 2011 *J. Magn. Magn. Mater.* **323** 651–68

[108] Kalambur V S, Han B, Hammer B E, Shield T W and Bischof J C 2005 *Nanotechnology* **16** 1221–33

[109] Xie J, Xu C, Kohler N, Hou Y and Sun S 2007 *Adv. Mater.* **19** 3163–6

[110] Hao R, Xing R, Xu Z, Hou Y, Gao S and Sun S 2010 *Adv. Mater.* **22** 2729–42

[111] Veiseh O, Gunn J W and Zhang M 2010 *Adv. Drug Delivery Rev.* **62** 284–304

[112] Gratton S E, Ropp P A, Pohlhaus P D, Luft J C, Madden V J, Napier M E and DeSimone J M 2008 *Proc. Natl. Acad. Sci. USA* **105** 11613–8
[113] Yallapu M M, Foy S P, Jain T K and Labhasetwar V 2010 *Pharm. Res.* **27** 2283–95
[114] Chen S, Li Y, Guo C, Wang J, Ma J, Liang X, Yang L R and Liu H Z 2007 *Langmuir* **23** 12669–76
[115] Gordon R T, Hines J R and Gordon D 1979 *Med. Hypotheses* **5** 83–102
[116] Rosensweig R E 2002 Heating magnetic fluid with alternating magnetic field *J. Magn. Magn. Mater.* **252** 370–4
[117] Andra W 1998 *Magnetism in Medicine: A Handbook* ed W Andra and H Nowak (Berlin: Wiley), 455
[118] Jeyadevan B 2010 Present status and prospects of magnetite nanoparticles-based hyperthermia *J. Ceram. Soc. Jpn.* **118** 391–401
[119] Dutz S and Hergt R 2013 Magnetic nanoparticle heating and heat transfer on a microscale: Basic principles, realities and physical limitations of hyperthermia for tumour therapy *Int J Hyperthermia* **29** 790–800
[120] Mamiya H 2013 Recent advances in understanding magnetic nanoparticles in AC magnetic fields and optimal design for targeted hyperthermia *J. Nanomater.* **2013** 752973
[121] Huang H S and Hainfeld J F 2013 Intravenous magnetic nanoparticle cancer hyperthermia *Int. J. Nanomed.* **8** 2521–32
[122] Yamada K *et al* 2010 Minimally required heat doses for various tumor sizes in induction heating cancer therapy determined by computer simulation using experimental data *Int. J. Hyperthermia* **26** 46
[123] Guidelines for limiting exposure to time-varying electric, magnetic, and electromagnetic fields (up to 300 GHz) 1998 *Health Phys.* **74** 494–522
[124] Döessel O and Bohnert J 2013 Safety considerations for magnetic fields of 10 mT to 100 mT amplitude in the frequency range of 10 kHz to 100 kHz for magnetic particle imaging *Biomed. Tech.* **58** 611–21
[125] Mamiya H, Takeda Y, Naka T, Kawazoe N, Chen G and Jeyadevan B 2017 Practical solution for effective whole-body magnetic fluid hyperthermia treatment *J. Nanomater.* **2017** 1047697
[126] Herman D A J, Ferguson P and Cheong S *et al* 2011 Hot-injection synthesis of iron/iron oxide core/shell nanoparticles for T 2 contrast enhancement in magnetic resonance imaging *Chem. Commun.* **47** 9221–3
[127] Qiang Y, Antony J, Sharma A, Nutting J, Sikes D and Meyer D 2006 Iron/iron oxide core–shell nanoclusters for biomedical applications *J. Nanopart. Res.* **8** 489–96
[128] Mamiya H and Jeyadevan B 2011 Hyperthermic effects of dissipative structures of magnetic nanoparticles in large alternating magnetic fields *Sci. Rep.* **1** 157
[129] Mamiya H and Jeyadevan B 2014 Magnetic hysteresis loop in a superparamagnetic state *IEEE Trans. Magn.* **50** 4001604

Chapter 8

Optical and magnetic detection of magnetic beads for medical diagnostics for point of care testing

Jaiyam Sharma, Kohki Noda, Ryoji Yukino, Tsukasa Takamura and Adarsh Sandhu

8.1 Introduction

Magnetic beads (MBs) offer many advantages as labels for biosensing, such as high surface-area-to-volume ratio, stable chemical and surface properties and the ability to be manipulated by magnetic fields. Thus, several technologies have been developed for biosensing based on the detection of MBs to measure the concentration of target biomolecules in a solution [1]. Sensing methods for the detection of MBs can broadly be classified into two groups: magnetic sensing and optical sensing. In this chapter, we begin with a review of the two major methods used for the detection of MBs based on magnetic and optical signals from MBs. Among the magnetic methods, giant magnetoresistance (GMR) sensors and Hall effect sensors are described.

The sensitivity of medical diagnostics platforms based on magnetic beads is limited by the amount of non-specific interactions between the MBs and sensing surface, as well as the sensitivity of detection of the MBs. Detection of magnetic beads by magnetic methods such as Hall effect sensors [2, 3] and giant magneto-resistance (GMR) sensors has been well studied [4, 5]. The sensitivity of detection of MBs by magnetic sensors is limited primarily due to two factors:

1. The strength of the magnetic field from MBs decreases as the third power of distance between the sensor and the MBs, i.e. magnetic field, $B = C/d^3$, where d is the distance from the sensor to the MB and C is a constant proportional to the volume of the MB. As a result, detection of nanometer sized MBs is quite challenging.
2. Since magnetic fields are used for the detection of MBs as well as their manipulation, the detection scheme suffers from noise due to the presence of the magnetic field used for the manipulation of MBs.

doi:10.1088/978-0-7503-1584-5ch8

Thus, other methods for the sensing of MBs have gained attention in recent years, such as using acoustic waves [6] and changes in the refractive index of materials by MBs [7]. Optical detection of MBs has been widely studied in the literature [8, 9]. In this chapter, we review two methods based on guided mode resonance (GR) sensors [10] and optical transmittance [11] of light from self-assembled chains of magnetic beads in a rotating magnetic field. However, optical detection, particularly, optical imaging, becomes challenging as the size of MBs decreases to less than the wavelength of visible light. To overcome these limitations, fluorescent magnetic nanoparticles (f-MNPs) have been reported [12, 13]. After a review of the existing literature, we describe the basic forces that act on MBs in solution and the problem of non-specific interactions between a solid substrate and MBs. We then describe a novel biosensing protocol aimed at overcoming these limitations, which forms the mainstay of this chapter.

The biosensing platform described reduces non-specific interactions between the sensing surface and f-MNPs by integrating a micro-coil pattern with the sensing area. We describe the design of current line patterns which can reduce non-specific interactions and their experimental evaluation. We describe a sensing protocol in which detection is performed by fluorescence observations and the number of f-MNPs are quantified by digital image processing. Finally, we demonstrate quantitative detection of biotin targets on a sensing surface functionalized with streptavidin.

8.2 Review of magnetic particle detection methods

Since the motivation for using MBs as labels for biosensing applications has already been discussed earlier in this book, here we assume that the reader is familiar with the process of functionalization of biomolecules on MBs. The detection of biomolecules requires the detection of MBs attached to probe antibodies. In general, the reaction between antibodies and antigens can occur anywhere in the volume of the liquid containing the biomolecules. However, if the antibodies are functionalized on a solid substrate, the reaction area is constrained to the surface and therefore only the MBs on the substrate serve as a signal for the biomolecules in solution. If, however, the antibodies are well dispersed within the solution, detection of MBs in the solution is required. This section covers both types of biosensing protocols.

8.2.1 Magnetic detection

In this section, we describe magnetic detection schemes for the detection of magnetic beads. We discuss various sensors based on the phenomena of giant magneto-resistance (GMR), tunnel magnetoresistance (TMR) and Hall effect.

When a current is passed through a conductor and an external magnetic field is applied in a direction perpendicular to the flow of current, a transverse voltage is developed across the conductor in a direction perpendicular to both the flow of current and the applied external magnetic field. This phenomenon is the well-known Hall effect, which was discovered by Edwin Hall in 1879. Hall effect sensors are widely used in industrial applications such as in rotary encoders of DC motors as well as in smartphones.

Sandhu *et al* [2] developed micro-Hall sensors for the detection and quantification of micro-sized magnetic beads. An alternating magnetic field was applied horizontally and a DC magnetic field was applied vertically (in the *z*-direction). Notably, the reported Hall sensor achieved very high sensitivity and detection of a single 2.8 μm magnetic particle was reported (figure 8.1).

GMR and TMR sensors are used in magnetic recording for hard disk drives (HDDs), and also have the potential to be used in the biomedical field [1]. Certain materials exhibit a change in their electrical resistance when exposed to an external magnetic field. This phenomenon is called the magnetoresistive effect. This phenomenon was first discovered and reported by W Thompson in 1857 [14]. For larger magnetoresistance change, the giant magnetoresistance effect was discovered by M N Baibich and co-workers in 1988 [15]. For thin films of certain materials, such as $Ni_{80}Fe_{20}$ (permalloy), NiFeCo and FeCr the ratio of change in resistance to the original resistance ($\Delta R/R$) is very high and thus, they are used as magnetic field sensors. In 1998, Baselt *et al* [16] demonstrated the use of magnetoresistive technology for the detection of magnetic beads. GMR sensors detect the presence of magnetic beads via a change in sensor resistance at a certain current due to the magnetic field generated by MBs.

Figure 8.2 shows an overview of various magnetic sensors versus their sensitivity [17]. Currently, the superconducting quantum interference device (SQUID) has the highest sensitivity for diagnosing the magnetic fields of both the brain and heart, using magnetoencephalography (MEG) and magnetocardiography (MCG), respectively. However, the SQUID needs a sophisticated superconducting device. GMR and TMR sensors are currently used for HDDs, automobiles, smartphones, and automated lines, and have potential for use in the biomedical field, as substitutes for the SQUID. In the near future, for instance, GMR and TMR sensors would potentially cost less, use less power, and would result in a device of simpler structure.

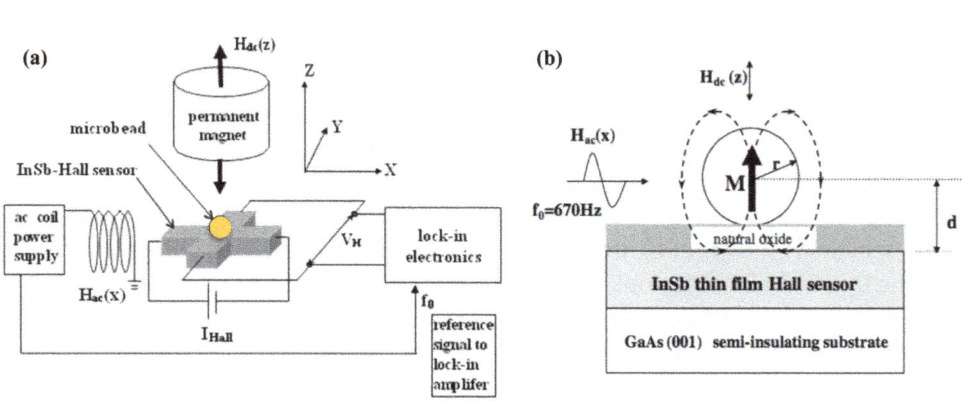

Figure 8.1. Schematic of a micro-Hall biosensor. (a) Experimental setup consisting of two sources of perpendicular magnetic fields. The magnetic field in the *x*-direction is AC whereas in the *z*-direction it is DC. (b) Schematic of magnetic field due to a magnetic particle on the sensing surface of a Hall sensor.

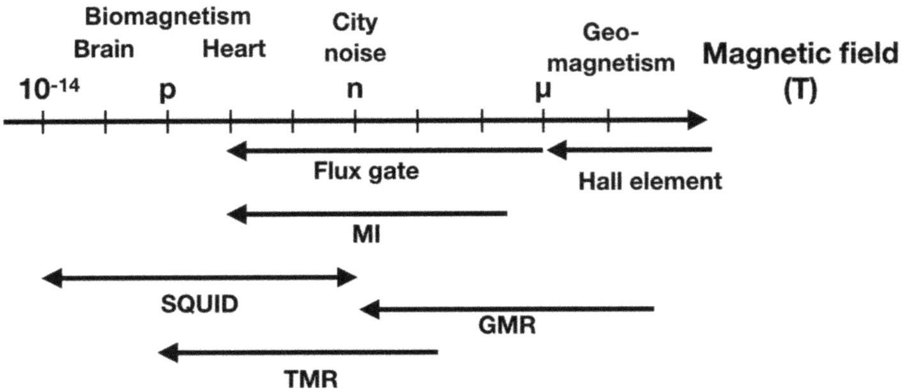

Figure 8.2. Various magnetic sensors versus their sensitivity.

Figure 8.3 shows the operating principle of a GMR sensor [18]. The basic structure consists mainly of three layers: a magnetic materials layer, a non-magnetic materials layer, and another magnetic materials layer. The upper magnetic layer, called the 'free layer', changes the direction of the magnetic moment according to the direction of the magnetic field applied. The middle layer below the free layer, called the 'pin layer', exchange-couples with the bottom layer, called the 'antiferromagnetic layer'. By this means of exchange-coupling, the direction of the pin layer is fixed with the direction of the antiferromagnetic layer. In addition, a thin insulating layer exists between the free layer and the pin layer, which creates a magnetic tunnel junction (MTJ). When the magnetic direction of both the free layer and the pin layer is parallel (the layers face the same direction), the magneto-resistance (MR) is low. The MR is high when the direction of both layers is antiparallel (the layers face the opposite direction), resulting in two states called 0 (low) and 1 (high), respectively. Thus, the MR ratio is an important parameter for the sensitivity of the sensor. In recent studies on MTJ for TMR [19], an MR ratio of 180% with a 500 mV voltage output have been recorded. The MTJ of the studies consists of Fe (001)/MgO (001)/Fe (001).

In practical applications, GMR sensors suffer from many sources of noise, such as pink noise (that varies with frequency as ~$1/f$), which is inherent in GMR sensors, as well as drift in voltage due to changes in temperature. Thus, GMR sensors are generally used in the configuration of a Wheatstone bridge as shown in figure 8.4 to minimize the effects of drift. A schematic of the detection of biomolecules with GMR sensors is shown in figure 8.5. Antibodies are functionalized on the surface of the sensor and MBs functionalized with antigens are brought in contact with the sensing surface by application of external magnetic fields. MBs containing antigens on their surface get bound to the sensing area due to antibody–antigen interactions. These MBs are then detected by GMR sensors and the concentration of biomolecules is then determined. In addition to GMR sensors, TMR sensors for biomedical applications has been developed and recently reported by Y Ando [17]. The first key feature of the sensor is a Wheatstone bridge circuit made with four TMR sensors.

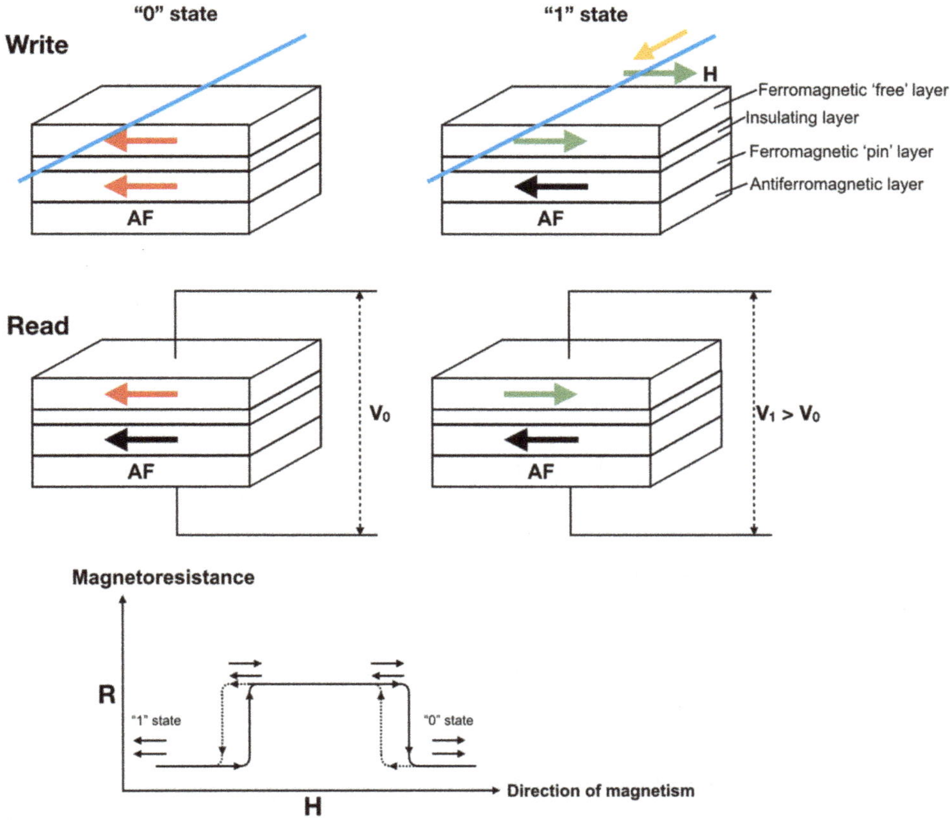

Figure 8.3. The operating principle of a GMR sensor.

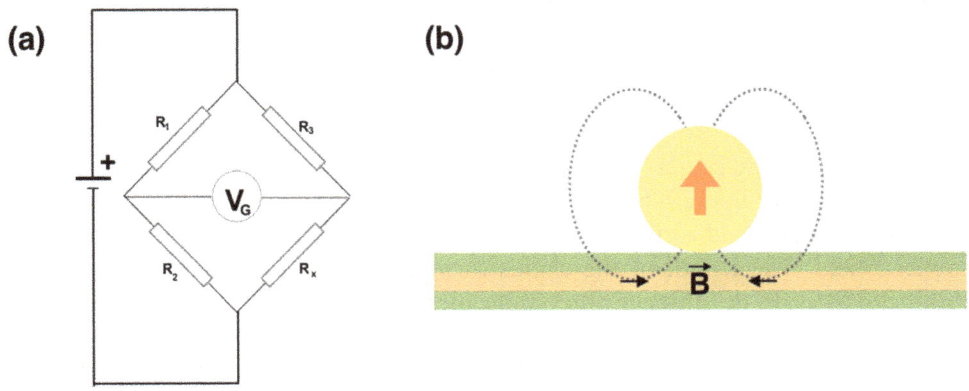

Figure 8.4. Principle of detection of magnetic nanoparticles with a GMR sensor. (a) Wheatstone bridge configuration used to minimize the effects of voltage drift, and (b) schematic of a magnetic field induced by a magnetic bead in the sensor.

The second feature is an amplifier with a filter circuit which reduces noise to the lowest possible voltage, especially reducing $1/f$ noise. By using this new type of sensor, the heart magnetic field signal has been detected; its data is comparable to that of SQUID, making this study very encouraging.

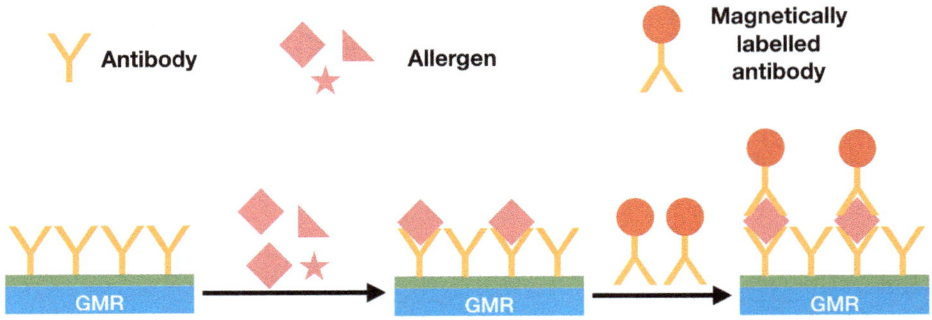

Figure 8.5. Schematic of the detection of biomolecules with giant magnetoresistance biosensors.

The magneto-impedance (MI) sensor has a different operating principle from those of the GMR/TMR sensors. The effect of magneto-impedance is the following: when an input pulse signal with a high frequency of more than several tens-MHz is sent through amorphous magnetic wires, the impedance of the wire changes according to the strength of the outside magnetic field. The output signal is received from the pick-up coil of the wire [20]. The MI has the potential to detect the femto Tesla (10^{-15} T) strength at room temperature, making it an effective substitute for the SQUID.

8.2.2 Optical detection of magnetic beads

Optical detection of magnetic beads offers a lot of promise, particularly from the point of view of utilizing such detection methods for point of care testing (POCT) applications. In this section, we review two notable efforts in this direction.

8.2.2.1 Magneto-optical biosensing platform

Park *et al* [11] have demonstrated a magneto-optical biosensing platform based on the scattering of light from self-assembled chains of magnetic beads in a rotating magnetic field. In the presence of an external magnetic field, MBs self-assemble into chains as shown in figure 8.6 (video).

As the magnetic field is rotated, the chains rotate to align with the direction of the magnetic field. The length of the chains formed depends on the interaction force among the MBs. In the absence of biomolecules in the solution, this interaction force is only due to magnetic interaction between the MBs. If the surface of the MBs is functionalized with antigens, and antibodies are added to the solution, the interaction between MBs increases due to antibody–antigen interactions. This results in an increase in the average length of the chains under the same magnetic field. This increase can be detected by measuring the transmittance of light through the solution. A schematic of the sensing principle is shown in figure 8.7. The orientation of chains follows the orientation of the external magnetic field. When the field is oriented perpendicular to the direction of the light source, the transmittance of light, measured by a photo-detector placed opposite to the light source, is minimum (figure 8.7(a)). As the field is rotated, the transmittance increases as the relative angle of orientation between the light source and magnetic field decreases (figure 8.7(b)). When the magnetic field is aligned with the light source, the transmitted light is maximum (figure 8.7(c)). The experimental results showed that a limit of detection of 100 pM was obtained with this method for the detection of avidin in solution.

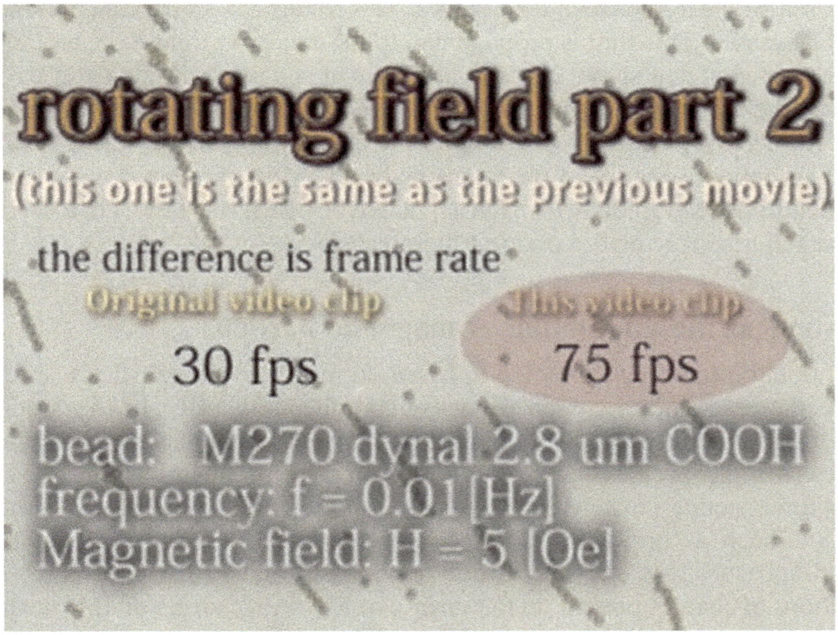

Figure 8.6. Video showing self-assembled chains of magnetic beads in a rotating magnetic field. Animation available from https://doi.org/10.1088/978-0-7503-1584-5.

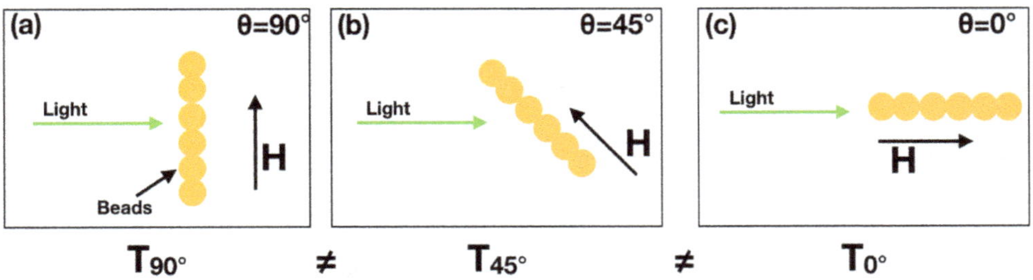

Figure 8.7. Schematic of a magneto-optical biosensing platform. The transmittance of light from the solution varies with the relative orientation of the external magnetic field with respect to the light source. (a) When the direction of the magnetic field is perpendicular to the light source ($\theta = 90°$), the transmitted light is minimum. (b) and (c) Transmittance increases as the relative angle decreases and is maximum at $\theta = 0°$. Adapted from Park *et al* [11].

8.2.2.2 Guided mode resonance based biosensor

Another notable development in the optical detection of magnetic particles has been reported by R Yukino *et al* [10], who demonstrated a guided mode resonance (GR) sensor consisting of silicon nitride thin film sub-wavelength nano-grating. When irradiated with broadband light, the reflection spectrum of the GR grating shows a characteristic peak whose position depends on the refractive index of the surface of the sensor. When biomolecules, such as avidin, are functionalized on the surface of the GR sensor, its refractive index changes. This results in a shift in the peak position of the GR sensor. This shift is proportional to the concentration of the biomolecules on the surface. They show that the addition of magnetic beads to the solution enhances the sensitivity of detection of the GR sensor by a factor of four. The enhanced sensitivity occurs because of an increase in the peak wavelength shift due to the high refractive index of the magnetic beads. Moreover, since magnetic beads can be manipulated rapidly with external magnetic fields, this results in a decrease in the time required for the detection of biomolecules.

8.3 Non-specific interaction of magnetic particles

Conventional biosensing technologies based on magnetic nanoparticles suffer from non-specific interaction between the sensing surface and MBs. In this section, we review the motion of MBs in a liquid solution under the influence of various forces and describe the problem of non-specific interactions quantitatively. There are various forces acting on an MB suspended in solution above a substrate as shown in figure 8.8, such as gravitational force, van der Waal's force between the MB's material and substrate, drag force due to the fluid and electrostatic force due to interactions of charges present on the surface of MB with charges on the substrate. If an external magnetic field is applied, an additional magnetic force is also present, whose magnitude and direction can be controlled through a suitable design of the field application method. For a small separation, z, between the MB and substrate, van der Waal's interactions become significant and can cause non-specific interactions between the MBs and substrate. The potential of the van der Waals forces between a sphere and a plane is given by

$$U_{\mathrm{vdw}} = \frac{A_{132}R}{6}[b \ln(\frac{1+bz}{bz}) - \frac{1}{z}]$$

where, A_{132} is the Hamaker constant of a sphere of material 1 on a substrate of material 2 in a medium of material 3. For the usual case of a polymer-coated MB on a silicon dioxide substrate in water, this constant is found to be 3.4 x 10^{-21} J, R is the radius of the sphere and $b = 14/\lambda_{\mathrm{ret}}$, where λ_{ret} is a characteristic length, generally assumed to be 100 nm. Figure 8.9 shows the van der Waals potential relative to thermal energy ($U_{\mathrm{vdw}}/k_{\mathrm{B}}T$) at room temperature for two commonly used MBs, Dynabeads and FG beads with diameters 2.8 μm and 180 nm, respectively. In the absence of an external magnetic field, van der Waal's forces become important when their magnitude is comparable to the thermal energy. Assuming a threshold of one order of magnitude for comparing the thermal energy and van der Waal's potential,

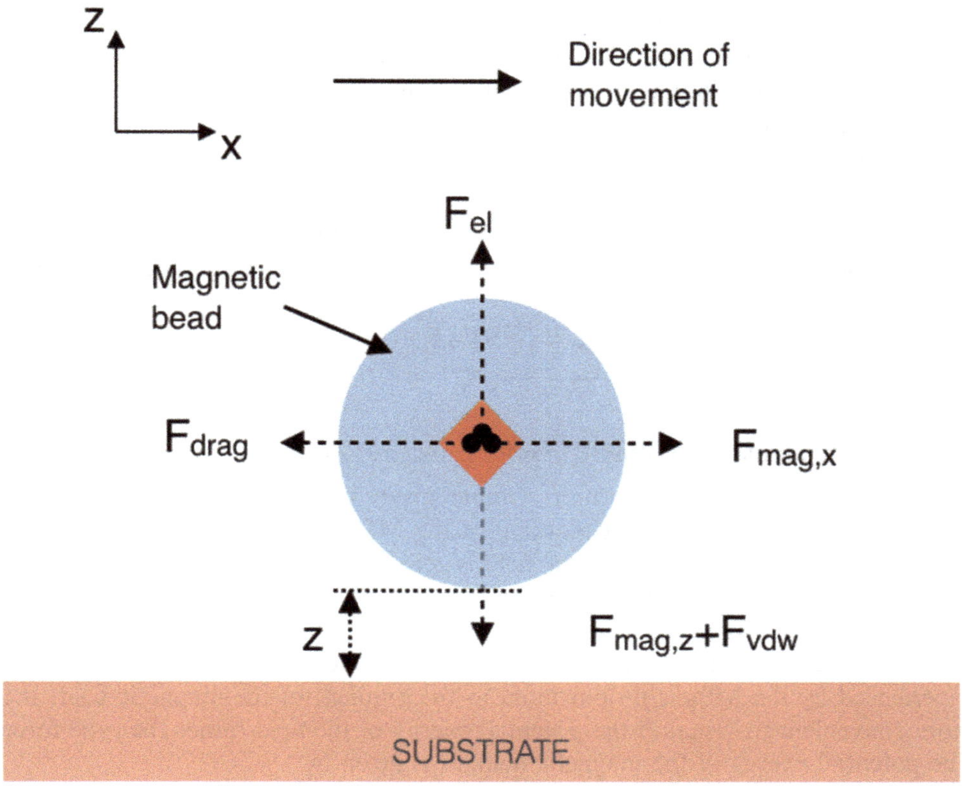

Figure 8.8. Schematic of forces acting on a MB suspended in solution.

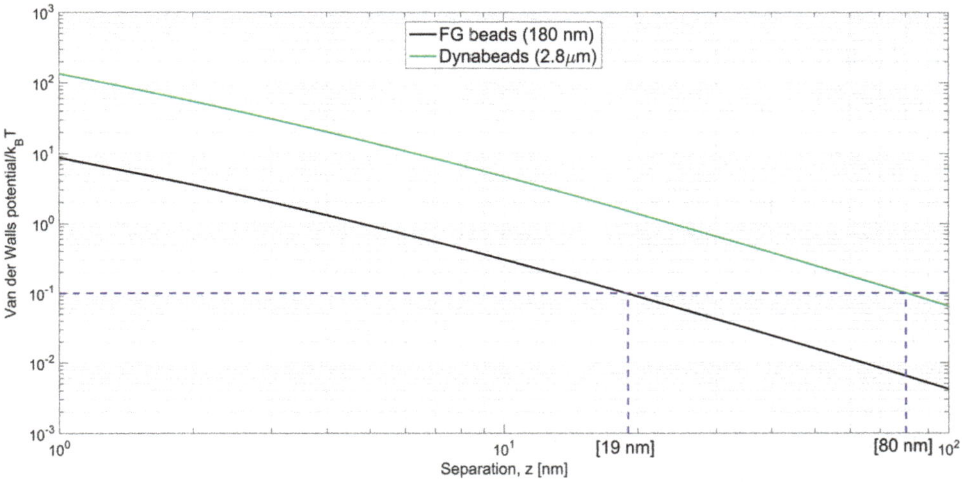

Figure 8.9. Potential of van der Waals interactions between MBs and substrate relative to thermal energy at room temperature (300 K).

we find that for typical nanometer sized MBs, van der Waal's forces become significant for a separation of about 20 nm, whereas for micrometer sized MBs, the corresponding separation is about 80 nm. Thus, methods for reduction of non-specific interactions should be effective at distances close to the substrate, typically less than 0.1 μm.

In terms of magnetism, a magnetic bead is a magnetic dipole consisting of a north and south pole. A magnetic dipole $\overrightarrow{m}$ placed under an external magnetic field $\overrightarrow{B}$ experiences a force $\overrightarrow{F_m}$ as well as a torque $\overrightarrow{\tau_m}$, mathematically expressed as [21]:

$$\begin{aligned} \overrightarrow{F_m} &= (\overrightarrow{m} \bullet \overrightarrow{\nabla})\overrightarrow{B} \\ \overrightarrow{\tau_m} &= \overrightarrow{m} \times \overrightarrow{B} \end{aligned} \tag{8.1}$$

The magnetic moment is given by $\overrightarrow{m} = V_b\overrightarrow{M}$, where V_b is the volume of the MB and $\overrightarrow{M}$ is the magnetization per unit volume given by $\overrightarrow{M} = \chi\overrightarrow{H}$, where χ is the susceptibility of the constituent material of the MB. If the external magnetic field is constant (i.e. gradient of magnetic field is zero), only torque is experienced. This causes the magnetic moments of MBs to be aligned with the applied magnetic field and causes the MBs to form chains. However, no motion of MBs occurs in a constant magnetic field. From equation (8.1), it follows that the magnetic force experienced by the MBs is proportional to the gradient of the magnetic field. It is more convenient to visualize the potential energy of the field rather than the force. The potential energy of the magnetic field, U is given by:

$$U = -\frac{V_b\chi}{2\mu_0}\left|\overrightarrow{B}\right|^2 \tag{8.2}$$

8.4 Design of micro-coils

Conventionally, magnetic assays use permanent magnets to induce interactions between the antibody functionalized sensing surface and antigen functionalized MBs [28]. Here, since the forces between the substrate and MBs are primarily vertical, in addition to the usually occurring non-specific interactions, magnetic forces can induce non-specific interactions as well. In order to avoid inducing non-specific interactions and expedite specific interactions on the sensing area, integrating micrometer sized current lines on the sensing area was proposed [22]. The use of on-chip 'magnetic stripes' for separation of biological cells tagged with MBs has also been reported [23]. Although MBs can be manipulated using permanent magnets, as well as micro-machined micro-magnets, manipulation with current line electrodes integrated on the substrate offers the most precise localized control over the motion of MBs. Thus, current line patterns have been used, for example, to determine antibody–antigen interaction forces between avidin and biotin [24]. In this section, we show how a micro-coil pattern is more effective at eliminating non-specific interactions while at the same being more efficient for collection of MBs on the sensing area.

The key requirements from an electrode design for the elimination of non-specific interactions are three-fold [29]:

1. The collection of a large number of MBs on the sensing surface to enhance the probability of specific interactions;
2. Elimination of non-specific interactions on the sensing area; and
3. Achieving the above steps with a sufficiently low value of current to minimize the heating effects on the biological sample.

The collection of MBs on the sensing surface requires a large magnetic force far away from the substrate, whereas elimination on non-specific interactions requires a high magnetic force close to the substrate surface. We performed a comparison of magnetic force profiles for linear and circular current lines for Dynabeads®. The magnetic field profiles were calculated using the well-known Biot–Savart law,

$$\vec{B} = \frac{\mu_0}{4\pi} \int \frac{I\, \vec{dl} \times \vec{r}}{|\vec{r}|^3}$$

where μ_0 is the permeability of free space, $\vec{dl}$ is a differential element on the current line pattern, $\vec{r}$ is the vector from the element $\vec{dl}$ to a test point, and the integral is over the volume of the electrode pattern. The magnetic forces were calculated from the magnetic field profile using equation (8.1). Figure 8.10 shows the results of a comparison of magnetic forces for the two patterns. A current of 10 mA was assumed for an electrode with width 5 μm and thickness 200 nm for both cases. Thus, the differences arise only due to the shape of the patterns.

Due to the curvature in the coil pattern, the magnetic field extends farther than that of a linear current line. To attract MBs towards the sensing area, the magnetic force should be greater than a certain threshold so as to overcome other forces such as electrostatic force and Brownian motion. This threshold was experimentally determined to be 0.1 pN. The area around the current line where the magnetic force is greater than the threshold force defines the capture cross-section of the current line. The white boundary in figures 8.10(c) and (d) represent the boundary of the 0.1 pN threshold. The capture cross-section of the coil pattern is 55% more than the linear current line. Thus, the coil pattern is expected to collect more MBs and bring them towards the sensing area. For the elimination of non-specific interactions, forces produced close to the substrate surface at a height of few tenths of a micrometer are important. A comparison of magnetic forces close to the substrate surface is shown in figure 8.11.

Here, the maximum force produced by current lines in the plane at a given height is used for comparison. As seen in figure 8.11, due to its curvature the coil pattern produces 3–4 times more force for the same amount of current close to the surface of the substrate. The ratio of forces decreases with height and far away from the substrate (say, 10 μm), the coil pattern produces about 1.5 times the force of a linear

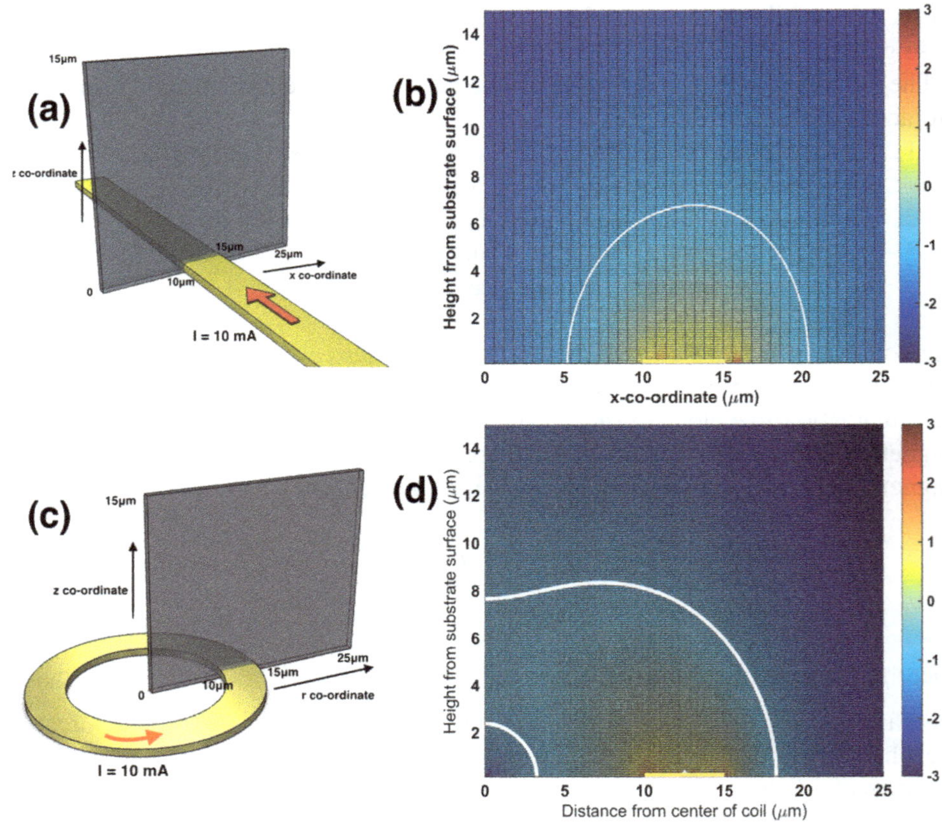

Figure 8.10. Comparison of magnetic force from linear and coil patterns. (a) and (c) Schematics of simulation regions for linear and coil patterns, respectively. (b) and (d) Magnitude of magnetic fields (in pN) for linear and coil patterns, respectively, on a logarithmic scale. The yellow bars show the location of the electrode in each case. The white boundary in each case represents the contour line for 0.1 pN force. The area within the boundary represents the capture cross-sections for the patterns. The circular coil has 55% more capture cross-section (124 μm^2) than a linear current line (80 μm^2).

electrode (data not shown). The ratio of forces should increase with increasing curvature of the coil (decreasing radius). Thus, utilizing coils that are as small as possible is preferable. However, from an experimental point of view there are three limitations that occur with decreasing coil size:

1. Micro-fabrication becomes difficult and produces low yields for sizes less than the resolution of micro-fabrication equipment and materials used. Most resolution equipment and materials (photo-resists) support a minimum resolution of 5 µm.
2. As the width of the electrodes decreases, electro-migration effects [26] become important. For a typical electrode material, such as gold, the maximum current density allowed through the material was experimentally determined to be ~2×10^6 A cm^{-2}. Thus, for an electrode of 5 µm and thickness 200 nm, the maximum current that can be applied is 10 mA.

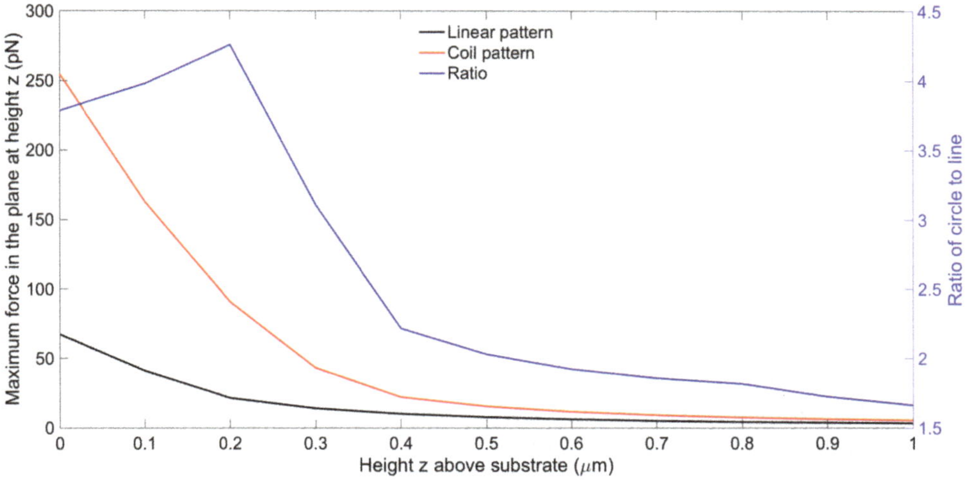

Figure 8.11. Comparison of magnetic forces produced in a plane by linear and coil patterns close to the substrate. This determines the ability to eliminate non-specific interactions. The force produced by a coil pattern is 3–4 times larger than that of a linear pattern close to the substrate (<0.2 μm). At very large heights the force produced by a coil pattern is 1.5 times that of a linear pattern.

3. As the size of an electrode decreases, its resistance increases due to increasing boundary effects. Thus, Joule heating occurs in the sample due to the flow of currents which leads to an increase in temperatures. High temperatures can lead to thermal denaturation of proteins [27], which adversely impacts specific interactions.

To summarize, in this section we have discussed the key points of consideration in designing micro-electrodes for the manipulation of MBs. Through electromagnetic simulation of magnetic fields, we showed that the coil pattern is more suitable for collecting MBs towards the sensing area as well as for the elimination of non-specific interactions. Finally, we discussed some experimental considerations for the implementation of electrode designs.

8.5 Experimental evaluation

In this section, we describe the experimental results obtained from a coil pattern. We show that the coil pattern is very effective for rapid manipulation of MBs on a substrate. Large micrometer sized MBs are used for demonstration purposes. Next, f-MNPs are used to demonstrate the elimination of non-specific interactions from the sensing area in a streptavidin–biotin model. Finally, we demonstrate quantitative detection of biotin concentration in a solution with a competitive assay.

A single coil is enough for the collection of MBs. However, for the elimination of non-specific interactions a second coil is required to enable the motion of MBs over the sensing area in both directions. A double coil pattern was designed as shown in the schematic in figure 8.12.

The annular region marked blue in the figure serves as the sensing area. MBs can be moved across the sensing area in both directions by alternating the coil in which the current is applied. The parameters of the coils such as the radii and widths determine the performance of the pattern. As remarked above, lower radius (higher curvature) coils are preferable at the cost of lower current and joule heating. Thus, the trade-off between size and performance is not clear from theoretical considerations. Therefore, several designs were fabricated for experimental evaluation.

8.5.1 Fabrication procedure

Current line patterns were fabricated using the following steps: (1) photolithography, (2) electron beam deposition of metals for making the electrodes, and (3) lift-off procedure to remove excess photoresist from the substrate. The fabrication procedure is summarized in figure 8.13.

First, a positive tone image reversal photoresist is spin-coated on a silicon dioxide wafer. Photolithography (i-line, 365 nm) is carried out with a photomask that contains the shape of the coil pattern to be fabricated. After a short exposure, the substrate is baked to allow an image reversal of the photoresist. After this process, the unexposed positive resist behaves as a negative resist. A flood exposure is carried out next. The wafer is then developed to expose the substrate where the electrodes are desired. Two layers of metal are deposited: 100 nm of titanium (Ti) and 100 nm

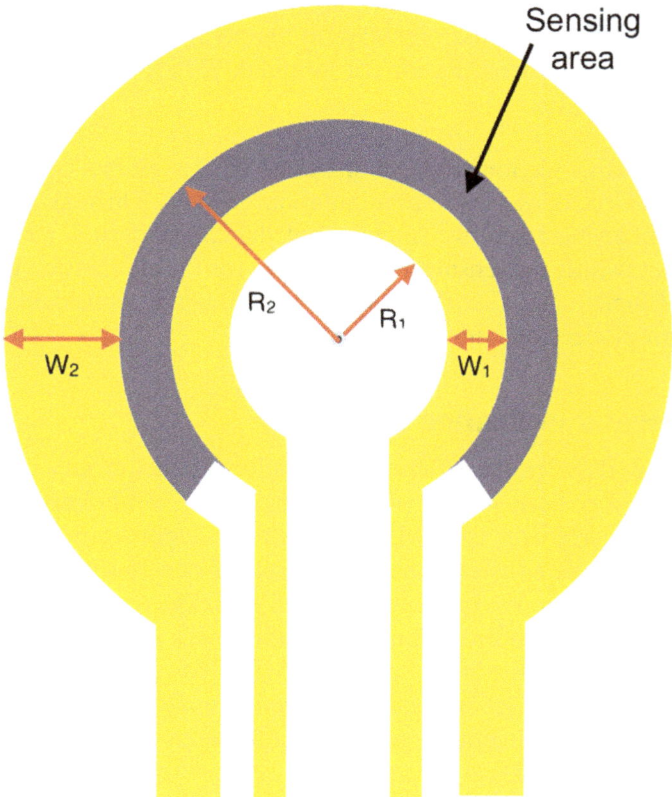

Figure 8.12. Schematic of an electrode pattern for the manipulation of MBs.

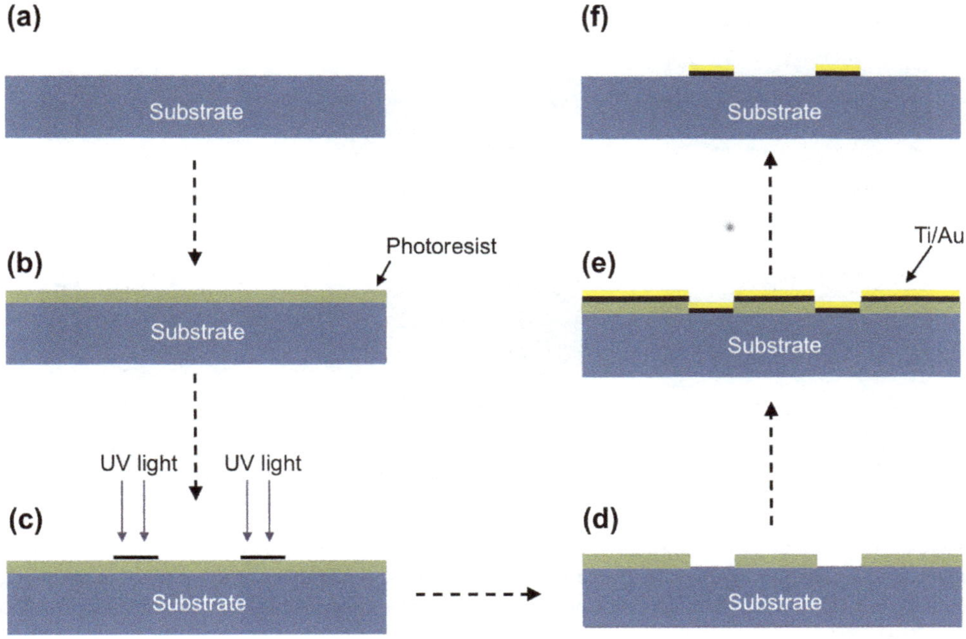

Figure 8.13. Schematic of the fabrication procedure for electrodes for micro-manipulation of MBs

of gold (Au), wherein the titanium layer acts as an adherent layer between the substrate and the gold. Finally, lift-off is carried out in an acetone solution with ultra-sonication to remove the excess photoresist from the substrate.

8.5.2 Manipulation of magnetic beads by micro-coils

Three coils were fabricated with the parameters (radius, width) = (10x μm, 5x μm) for $x = 1,2,3$. Thus, the coils were scaled replicas of one design. The magnetic force from the largest coil with a radius of 30 μm and width 15 μm was not enough to affect a motion in MBs within a few minutes (experimental data not shown). However, the coils corresponding to $x = 1$ and $x = 2$ (with radii 10 μm and 20 μm) show an excellent ability to collect and move MBs across the annular region between them, which is therefore designated as the sensing area. The sensing area is about 410 μm^2. In a subsequent discussion, the largest coil is ignored and results for the smaller two coils are shown. To compare the performance of the linear electrodes and coil pattern, experiments were carried out with 2.8 μm diameter dynabeads, as micrometer sized MBs can be easily observed with an optical microscope. Figure 8.14 shows a video of manipulation of the MBs across the sensing area in real time. An optical microscope (Keyence VH-450Z) was used for the observation of MBs. As shown, the MBs can be transported across the sensing area in a few seconds after the application of current. A comparison of the performance of linear and coil patterns can be made from these results. Figure 8.15 shows the quantitative performance of the fabricated patterns.

The surface was not functionalized with biomolecules for this experiment. A DC current of 10 mA was passed through the smallest coil for 30 s. As seen in figure

Figure 8.14. Real-time video showing the movement of 2.8 μm Dynabeads across the sensing area due to magnetic field generated by the coils. The current in the innermost coil is 10 mA (DC) and that in the middle coil is 20 mA (DC). The outermost coil is not connected. Animation available from https://doi.org/10.1088/978-0-7503-1584-5.

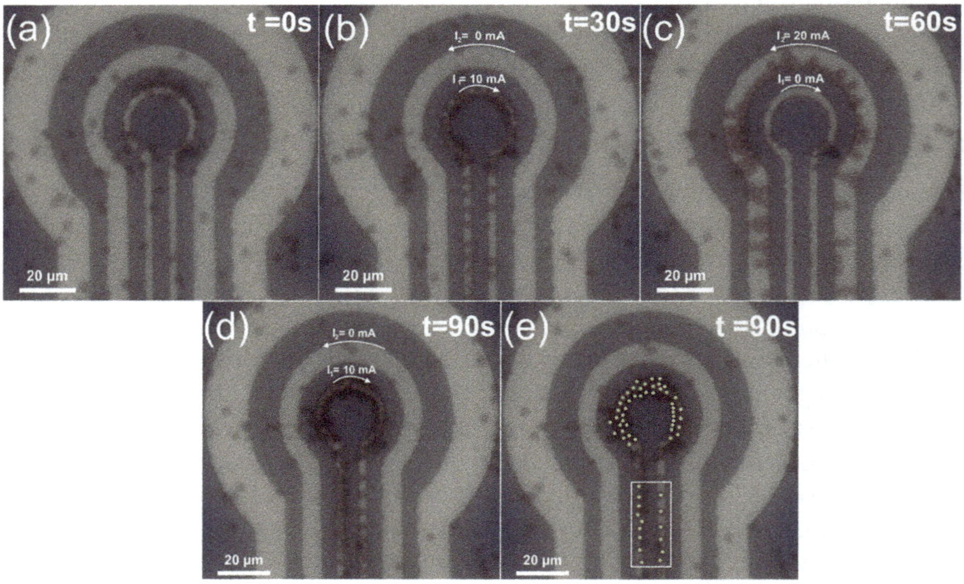

Figure 8.15. Demonstration of the manipulation of MBs by micro-coils. (a) Initial distribution of MBs, (b) MBs collect on the inner coil on the application of current, (c) MBs move across the coils by changing the coil where current is applied, (d) and (e) comparison of MBs collected on the coil pattern versus the linear pattern after 90 s. The linear electrodes within the white box shown in (e) have the same surface area as the coil electrode. However, the coil collects more than three times as many MBs (shown with green circles) as the linear pattern.

8.15(b), MBs collect over the coil. The MBs are moved across the sensing area by applying a current of 20 mA to the outer coil for 30 s, as shown in figure 8.15(c). This process can be repeated as shown in figure 8.15(d). The performance of the linear and coil patterns is compared in figure 8.15(e). We compare the number of MBs collected by electrodes with the same surface area but different shapes. The linear electrodes in the white box have the same surface area as the inner coil. The MBs are marked with green circles for visual aid. For the same current, the circular coil collects about three times as many MBs (47 MBs) as the linear pattern (15 MBs). This experiment was repeated many times and the ratio of MBs collected by the coil pattern to linear pattern was consistently found to be between three and four. It is important to note that the performance of the linear electrodes was adversely affected due to the opposite direction of the current between the left and right linear electrodes, which led to opposite directions of magnetic fields and reduced the forces. Thus, the apparently large difference in the performance of the two patterns is due to both factors: shape as well as the direction of fields. In the experiment presented in figure 8.15, there were no biomolecules on the surface and thus no specific interactions occurred. However, if the surface is functionalized with probe molecules and the MBs are functionalized with corresponding target molecules, repeated motion of MBs across the sensing area will increase the probability of specific interactions, while eliminating non-specific interactions. This can reduce the time required for performing the diagnostics. From these results, we conclude that the double coil pattern with coils of radii 10 μm and 20 μm can successfully manipulate MBs across the sensing area. This was also found to be true for f-MNPs with a diameter of 180 nm.

8.5.3 Elimination of non-specific interactions

Next, we show the elimination of non-specific interactions with streptavidin and biotin as probe and target molecules, respectively. In this experiment, the substrate surface was functionalized with streptavidin and f-MNPs (FG beads, diameter 180 nm) were functionalized with biotin. The f-MNPs were imaged with a fluorescent microscope (Olympus BX-51). During the experiment, the current was increased in steps with time to the maximum value. Thus, the f-MNPs moved slowly across the sensing area. This is necessary since specific interactions occur over the timescale of at least a few seconds and very rapid motion of MBs may prevent specific interactions. The results are shown in figure 8.16. Figure 8.16(a) shows the profile of the current with time. The current in the inner coil is increased in steps of 1 mA from 0 to 10 mA over 60 s and is kept constant at 10 mA for the next 60 s. Thus, f-MNPs move once across the sensing area. This current is then switched off and applied to the outer coil. In this case, the current is increased in steps of 2 mA up to 20 mA over 60 s and held constant for another 60 s. Thus, the whole process takes only 4 min. As shown in figure 8.16(b), initially the sensing area (marked with a green boundary) contains very few f-MNPs. After the application of currents, as described above, a large number of f-MNPs are observed in the sensing area due to specific interactions as shown in figure 8.16(c). Figure 8.16(d) shows a false color profile of the sensing area. Areas estimated to contain f-MNPs are labeled pink, whereas areas estimated not to contain f-MNPs are labeled black. This profile enables a quantitative estimation of the number of f-MNPs over the sensing area. For this experiment, it was determined that approximately 68% of the sensing area was covered with f-MNPs. The number of f-MNPs was estimated to be about 1.4×10^4. However, the number of streptavidin–biotin specific interactions on the sensing area can be much larger than this number since each f-MNP is functionalized with a few thousand biotin molecules on its surface. The method for estimating the number of f-MNPs from images is described in section 8.7. It is worth noting that the f-MNPs in the sensing area do not move to the electrodes even though a magnetic force is applied to them by the electrodes. This confirms the presence of specific interactions between the f-MNPs and the substrate. In control experiments, where the surface was not functionalized with streptavidin,

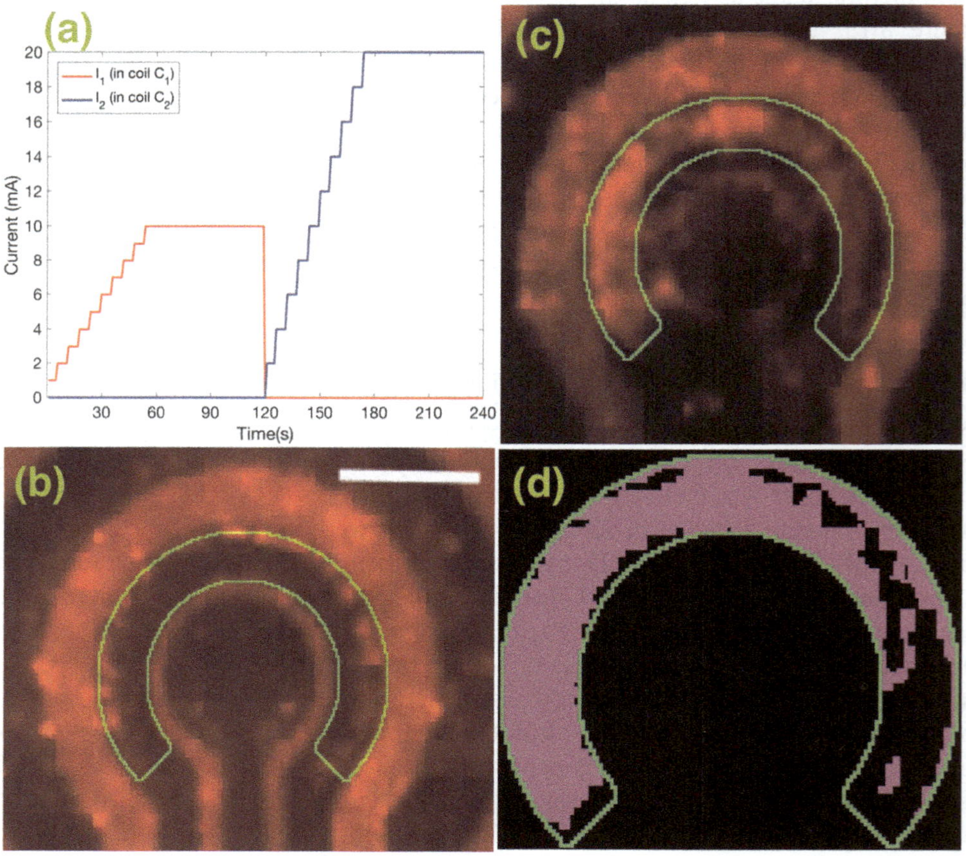

Figure 8.16. Demonstration of the elimination of non-specific interactions on a surface functionalized with streptavidin. (a) Current applied to coils with time, (b) initial state of the sensing area (within the green boundary) with very few f-MNPs, (c) sensing area after the application of current. A large number of f-MNPs can be observed. (d) False color profile of the sensing area shown in (c). Areas estimated to contain f-MNPs are colored pink. The scale bars are 20 μm for both (b) and (c).

biotin coated particles did not interact with the sensing area and could be easily transported over to the electrodes. The results from the control experiment are shown in figure 8.17.

8.6 Quantitative detection of biotin

Quantitative detection of biotin can be performed using double micro-coils. In this series of experiments, a quantitative assay method was used. Streptavidin probes were functionalized on the substrate. As above, FG beads were used as fluorescent magnetic nanoparticles. There were two sources of target molecules of biotin: f-MNPs functionalized with biotin as well as free biotin present within the solution. Thus, during the detection procedure, there was competition among the two types of biotin to interact with the streptavidin antibodies on the substrate. The concentration of biotin on the surface of f-MNPs was the same in all experiments and only the concentration of free biotin was varied from 1 nM (one nanomoles per liter) to 100 μM. Free biotin molecules (with a size of about 5 nm [30]) are much smaller than f-MNPs and exhibit more Brownian motion. Thus, free biotin may interact preferentially with the streptavidin on the surface before f-MNPs have a chance. Thus, one expects a lower number of f-MNPs on the substrate when the concentration of free biotin in the solution is high and vice versa.

In this series of experiments, currents were applied for 20 min. Cycles of 4 min as shown in figure 8.16(a) were repeated five times and a video of the sensing area was recorded with a fluorescent microscope. This was used for analysis to determine the percentage of the sensing area that is covered by f-MNPs. The results are shown in figure 8.18. Five different concentrations of free biotin (100 μM, 10 μM, 1 μM, 100 nM and 1 nM for figures 8.18(a)–(e), respectively) were used.

As expected, the number of f-MNPs on the sensing area increases with decreasing concentration of free biotin in the solution. The estimated percentage of the sensing area covered by f-MNPs and the estimated number of f-MNPs for all cases is shown in figure 8.19.

To summarize, in this section we showed the quantitative detection of free biotin in solution based on a competitive assay method using micro-coils for the reduction of non-specific interactions with the sensing surface.

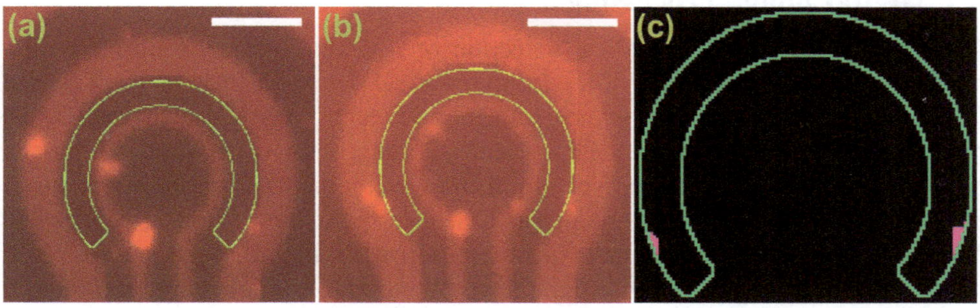

Figure 8.17. Control experiment for the elimination of non-specific interactions: (a) Initial state of the sensing area (within the green boundary), (b) after applying the currents as shown in figure 8.16(a), (c) estimation of areas covered by f-MNPs in (b). The scale bars in (a) and (b) represent 20 μm.

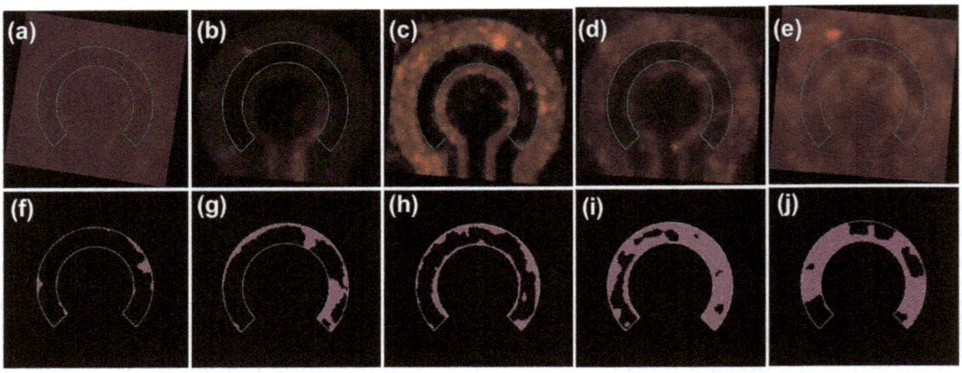

Figure 8.18. Results of the quantitative detection of free biotin in solution with micro-coils. (a)–(e) Show fluorescent images of the sensing area (marked green), (f)–(j) show the profile of coverage of f-MNPs on the sensing area. The areas estimated to be covered with f-MNPs are marked pink. The concentrations of free biotin in solution are (a) 100 μM, (b) 10 μM, (c) 1 μM, (d) 100 nM and (e) 1 nM.

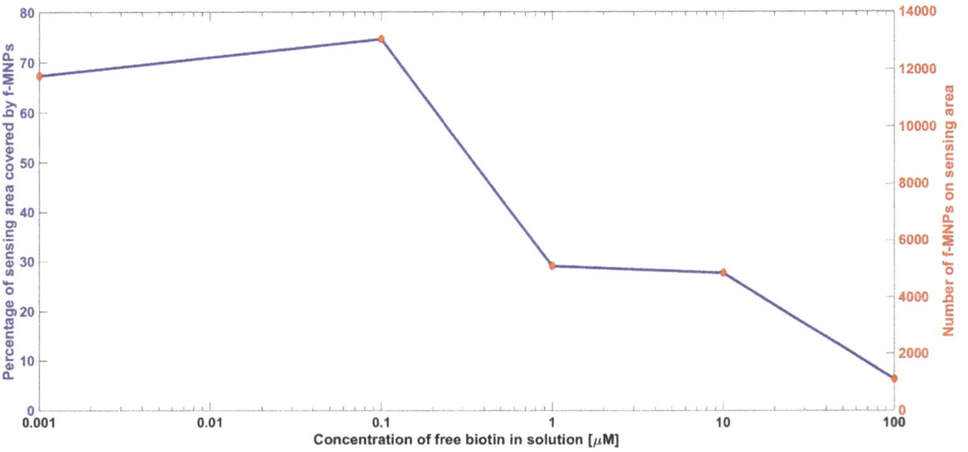

Figure 8.19. Calibration curve for the streptavidin–biotin complex obtained for the micro-coil biosensing platform.

8.7 Digital image processing

A significant advantage of optical detection of f-MNPs over conventionally used magnetic methods is the ease of integrating optical detection into compact devices for point of care diagnostics. Since smartphone technology has proliferated very rapidly into the world over the last decade, integration of medical diagnostics platforms with smartphones offers promising development. The diagnostic procedure described in the previous section is based on the observation of f-MNPs with fluorescence microscopy and their quantification with digital image processing. In this section, we describe a simple and effective method for quantitative estimation of f-MNPs on the sensing area with digital image processing of fluorescence microscopy images. We also discuss some advantages and limitations of the method described.

The estimation of the number of f-MNPs can be divided into two steps:

1. Localizing the sensing area within the image.
2. Identifying the regions within the sensing area which contain f-MNPs.

When there are just a few experiments to be analyzed, the steps mentioned above can be performed manually with input from humans. However, such systems cannot scale well with the number of samples to be analyzed. Thus, we describe a fully automated approach for carrying out the steps outlined above. More details about the computer vision algorithms mentioned in this section can be found in reference [31].

8.7.1 Localization of sensing area

In general, the sensing area can be anywhere in the image and is not limited to a specific region. The sensing area as shown in figure 8.12, is composed of the annular region between the two coils. Thus, in order to recognize the sensing area, the task is that of localizing the two coils. There are several ways to perform the localization. One simple approach involves using a Canny edge detector to find the edges in the image and then using the edges to identify the coils. This is generally accomplished using a Hough transform. Other approaches that may be used for performing the localization in the image are Viola–Jones object detectors, template matching and combinations of these approaches.

In the present work, we used the Hough transform for the task. The aspect ratio of objects to be detected is not fixed for Hough transforms as we only use them to identify primitive shapes such as lines and (in this case) circles. This is advantageous for our task since fabrication errors between substrates during photolithography and metal deposition processes can result in the sensing area of the fabricated substrates to vary from the design by more than 10%. To account for such differences, we use Hough transforms for detection of circles within the image and these results are combined to infer the location of the sensing area. Other methods, such as template matching and Viola–Jones detectors, detect the entire pattern in one step and cannot account for the differences between individual substrates. Additionally, we use Hough transforms to detect the orientation of the electrodes connecting the micro-coils, which helps take into account the rotation of the coils that occur due to experimental error in positioning the substrate. Thus, the localization step is quite general and can take errors during fabrication as well as experimentation into account.

8.7.2 Identification of regions containing f-MNPs

Once the location of the sensing area has been identified, the last step for the quantitative estimation of f-MNPs is to identify the regions of the sensing area that contain f-MNPs. This can be looked at as an image segmentation problem and various segmentation methods may be used to identify the f-MNPs. In the simplest approach, a binary segmentation method such as Otsu's method may be used. Other segmentation methods, such as graph cuts or methods for clustering such as k-means and Gaussian mixture models, may also be used to solve the segmentation problem. However, graph-cut methods require manual input from the user with an initial guess of the background and are non-deterministic, as is also the case for clustering methods. Non-deterministic algorithms produce different results each time when run on the same image. This is undesirable since it introduces an additional source of error in the estimation of f-MNPs in addition to sources of experimental error. Thus, in the present work we used Otsu's method for the determination of a threshold intensity between the substrate (dark regions in the image) and f-MNPs.

To obtain an unbiased estimate, we calculate the threshold for regions outside the sensing area. Importantly, the surfaces of gold micro-coils have a higher reflectivity than silicon dioxide substrate and will bias our estimate of the threshold. Therefore, the regions containing gold were not used for determination of the threshold. To ensure that the threshold intensity between the substrate and f-MNPs is constant across all locations in the image, we estimate the illumination profile of the incident light with morphological opening of the image. This produces an estimate of the background illumination, which takes non-uniform illumination into account. This background is then subtracted from the original image. A threshold is calculated on the resulting image. Regions in the sensing area where the intensity is greater than this threshold are deemed to contain f-MNPs and vice versa.

The two steps discussed above and some other minor processing steps are shown in figure 8.20. A video of the substrate is recorded with a fluorescent microscope. A high-resolution image is obtained from stacking several frames and enhancing the resulting intensity of the image by histogram stretching. Since the focused area is much smaller than the sample volume, the captured image contains light from outside the focused area. Thus, the resulting image contains high fluorescence intensity even in regions that do not contain f-MNPs. These effects and those due to non-uniform illumination from the light source are reduced by estimating the background illumination, that is the profile of light source if there were no coils on the substrate. This background is then subtracted from the original image. The location of the sensing area is detected by using the Hough transform as shown in figure 8.20(d). From figure 8.20(a), it is clear to see that the linear electrodes are slightly rotated from the vertical. This rotation is estimated by the orientation of the straight lines shown in green in figure 8.20(d). In this case the rotation was found to be about 5 degrees from the vertical. The region shown by the green box in figure 8.20(e) is used for estimation of the threshold intensity between the f-MNPs and substrate. Thus, the regions in the sensing area containing f-MNPs are estimated, as shown in figure 8.20(f). The estimation of the number of f-MNPs from the obtained profile of the sensing area requires some assumptions. The first assumption is that the f-MNPs are all located in a plane. In general, the f-MNPs may form multi-layered structures or three-dimensional clusters due to their ferromagnetic nature. However, only the f-MNPs that are attached to the sensing area are of interest to us. Thus, for the purpose of analysis it is assumed that all f-MNPs lie in a plane. Secondly, it is assumed that the f-MNPs self-assemble in a hexagonal arrangement. Self-assembly of magnetic beads in the presence of a magnetic field into planar hexagonal patterns has been reported by Y Yang *et al* [25]. Thus, the number of

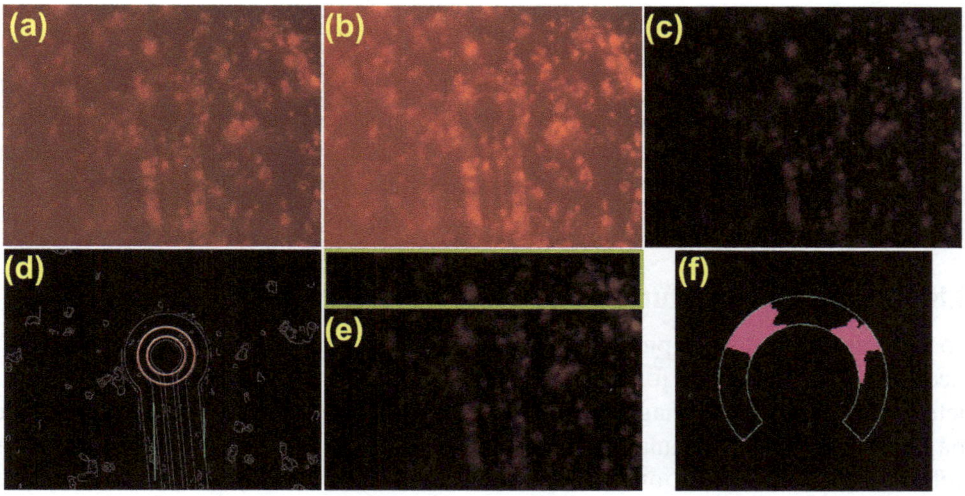

Figure 8.20. Procedure to calculate the amount of f-MNPs on the sensing area. (a) A video of the substrate is recorded from the microscope, (b) a high quality denoised image is obtained from stacking frames, stretching its histogram. Notice that the resulting image contains large illumination due to fluoresence from the sample volume outside the focused region. (c) These effects are reduced by the morphological opening of the image in (b). (d) Edge profile of the image including detected circles (red) and orientation of the sample measured by the angle of the detected lines (green). (e) The region excluding the coils marked by the green box is used for calculation of the threshold, (f) result from binary thresholding of the sensing area.

f-MNPs can be calculated by multiplying the total area covered by f-MNPs by the packing fraction of the hexagonal packing structure (91%) and dividing by the area covered by one magnetic nanoparticle. For the example shown in figure 8.20, it was estimated that about 5200 f-MNPs were present on the sensing area. It is important to note that the estimate of the absolute number of f-MNPs may be erroneous as one or both of the assumptions stated above may be violated in the experiments. However, this procedure is quite successful at estimating the trends of the number of f-MNPs in different experiments. Independent verification of the validity of this procedure by comparison with, for example, scanning electron microscopy images, was not carried out in this study. However, maps of the sensing area (similar to figure 8.20(f)) produced by this procedure agree well with visual inspection of the fluorescent images. Further refinement and calibration of this method with more accurate detection methods may be necessary before applying it in experimental conditions than the ones used in this study.

Some of the advantages of the method outlined above are:

1. The method accounts for the errors in fabrication of electrode patterns during photolithography and metal deposition processes.
2. No reference light is required during the experiment since the threshold between the substrate and f-MNPs is determined using other regions in the same image.

Some limitations of the method are:

1. The estimation of the absolute number of f-MNPs in the sensing area may be erroneous. In this study, only trends of number of f-MNPs were considered.
2. In cases where there are very few f-MNPs in the sensing area as well as outside the sensing area, the estimate of the threshold will be quite noisy and the estimated number of f-MNPs may be incorrect.

8.8 Conclusion and future work

To summarize, in this chapter we have presented an outline of a biosensing platform incorporating micro-coils for the elimination of non-specific interactions and optical detection of fluorescent magnetic nanoparticles. The magnetic forces acting on the magnetic beads were simulated and it was found that the coil pattern was better suited for both the collection of magnetic beads towards the sensing area as well as for the elimination of non-specific interactions. The coil patterns were fabricated on a silicon substrate using conventional photolithography technology followed by electron beam deposition of titanium and gold metals. Experiments showed that the coil pattern was indeed superior to the linear electrodes for attracting magnetic beads towards the sensing area. Next, we demonstrate a reduction of non-specific interactions using fluorescent magnetic nanoparticles. It was shown that f-MNPs that are attached to the sensing area by specific interactions cannot be removed on application of the current in the micro-coils. Subsequently, we showed quantitative detection of biotin with a competitive method.

Research efforts for biosensing with magnetic particles in the future are likely to focus on the development of small, portable and affordable protocols and systems for medical diagnostics. Since such systems may allow the possibility of diagnostic tests to be carried out by patients themselves, these systems are called point of care testing (POCT) systems, as opposed to conventional technology, which requires tests to be performed away from the point of care in medical laboratories.

References

[1] Sandhu A 2007 Biosensing: New probes offer much faster results *Nat. Nanotechnol* **2** 746

[2] Sandhu A, Sanbonsugi H, Shibasaki I, Abe M and Handa H 2004 High sensitivity InSb ultra-thin film micro-Hall sensors for bioscreening applications *Jpn. J. Appl. Phys. Part 2 Lett.* **43** 868–70

[3] Besse P, Boero G, Demierre M, Pott V and Popovic R 2002 Detection of a single magnetic microbead using a miniaturized silicon Hall sensor *Appl. Phys. Lett.* **80** 4199

[4] Graham D L, Ferreira H A and Freitas P P 2004 Magnetoresistive-based biosensors and biochips *Trends Biotechnol.* **22** 455–62

[5] Freitas P P *et al* 2012 Spintronic platforms for biomedical applications *Lab Chip* **12** 546

[6] Tano M, Nakagawa T, Ueda T, Seino S and Yamamoto T A 2013 Extremely long signal delays from magnetic particles *Mater. Lett.* **98** 51–4

[7] Bruls D M *et al* 2009 Rapid integrated biosensor for multiplexed immunoassays based on actuated magnetic nanoparticles *Lab Chip* **9** 3504

[8] Morimoto Y, Abe M, Hatakayama M, Handa H and Sandhu A 2009 Detection of magnetic nanobeads by self-assembly of superparamagnetic microbeads for biosensing *IEEE Trans. Magn.* **45** 2871–4

[9] Yang Y, Morimoto Y, Takamura T and Sandhu A 2012 Biosensing based on magnetically induced self-assembly of particles in magnetic colloids *J Nanosci Nanotechnol.* **12** 2081–8

[10] Yukino R, Sharma J, Takamura T, Joseph J and Sandhu A 2018 Magnetic nanoparticle-based nano-grating guided-mode resonance biosensors *IEEE Trans. Magn.* **54** 5100306

[11] Park S Y, Handa H and Sandhu A 2010 Magneto-optical biosensing platform based on light scattering from self-assembled chains of functionalized rotating magnetic beads *Nano Lett.* **10** 446–51

[12] Nishio K *et al* 2008 Development of novel magnetic nano-carriers for high-performance affinity purification *Colloids Surf.* B **64** 162–9

[13] Hatakeyama M, Mochizuki Y, Kita Y, Kishi H, Nishio K, Sakamoto S, Abe M and Handa H 2009 Characterization of a magnetic carrier encapsulating europium and ferrite nano-particles for biomolecular recognition and imaging *J. Magn. Magn. Mater.* **321** 1364–7

[14] Thompson W 1858 *Proc. R. Soc. London* **8** 546

[15] Baibich M N, Broto J M, Fert A, Nguyen Van Dau F, Petroff F, Etienne P, Creuzet G, Friederich A and Chazelas J 1988 *Phys. Rev. Lett.* **61** 2472

[16] Baselt D R *et al* 1998 A biosensor based on magnetoresistance technology *Biosens. Bioelectron.* **13.7** 731–9

[17] Ando Y 2017 *Proc., Iwasaki Conf. No. 5, MTJ, 17*

[18] Noda K 2006 *Biomedical and Environmental Applications of Functionalized Magnetic Beads* ed M Abe, H Handa and K Nodavol 120 The original in Japanese

[19] Yuasa S, Nagahama T, Fukushima A, Suzuki Y and Ando K 2004 *Nat. Mater.* **3** 868

[20] Uchiyama T 2016 *Proc., Iwasaki Conf. No. 4, MTJ, 43*

[21] Wirix-Speetjens Roel *et al* 2005 A force study of on-chip magnetic particle transport based on tapered conductors *IEEE Trans. Magn.* **41** 4128–33

[22] Kumagai Y, Togawa K, Sakamoto S, Abe M, Handa H and Sandhu A 2006 Hall biosensor with integrated current microstrips for control of magnetic beads *IEEE Trans. Magn.* **42** 3893–5

[23] Derks R J S *et al* 2007 Magnetic bead manipulation in a sub-microliter fluid volume applicable for biosensing *Microfluid. Nanofluid.* **3.2** 141–9

[24] Panhorst M *et al* 2005 Sensitive bondforce measurements of ligand–receptor pairs with magnetic beads *Biosens. Bioelectron.* **20.8** 1685–9

[25] Yang Y, Li Z, Ko P J and Sandhu A 2012 Behavior of multi-component magnetic colloidal systems in tunable magnetic fields and applications in biosensing *J. Phys. Conf. Ser.* **352** 12001

[26] Sze S M 2008 *Semiconductor Devices: Physics and Technology* (New York: Wiley)

[27] Wang X 2014 Characterization of denaturation and renaturation of DNA for DNA hybridization *Environ. Health Toxicol* 29

[28] Sandhu A, Handa H and Abe M 2010 Synthesis and applications of magnetic nanoparticles for biorecognition and point of care medical diagnostics *Nanotechnology* **21** 442001

[29] Sharma J, Ishizawa S, Yukino R, Takamura T, Hanyu N, Yasuno H and Sandhu A 2016 Fast and sensitive medical diagnostic protocol based on integrating circular current lines for magnetic washing and optical detection of fluorescent magnetic nanobeads *Sens. Bio-sens. Res.* **9** 7–12

[30] Livnah O, Bayer E A, Wilchek M and Sussman J L 1993 Three-dimensional structures of avidin and the avidin–biotin complex *Proc. Natl. Acad. Sci.* **90** 5076–80
[31] Szeliski R 2010 *Computer Vision: Algorithms and Applications* (Berlin: Springer)
[32] Millen R L, Kawaguchi T, Granger M C, Porter M D and Tondra M 2005 *Anal. Chem.* **77** 6581

Lightning Source UK Ltd.
Milton Keynes UK
UKHW05n1615231018
331042UK00004BB/140/P

9 780750 315821